METHODOLOGY AND TECHNOLOGY
OF DECOMMISSIONING NUCLEAR FACILITIES

The following States are Members of the International Atomic Energy Agency:

AFGHANISTAN	GUATEMALA	PARAGUAY
ALBANIA	HAITI	PERU
ALGERIA	HOLY SEE	PHILIPPINES
ARGENTINA	HUNGARY	POLAND
AUSTRALIA	ICELAND	PORTUGAL
AUSTRIA	INDIA	QATAR
BANGLADESH	INDONESIA	ROMANIA
BELGIUM	IRAN, ISLAMIC REPUBLIC OF	SAUDI ARABIA
BOLIVIA	IRAQ	SENEGAL
BRAZIL	IRELAND	SIERRA LEONE
BULGARIA	ISRAEL	SINGAPORE
BURMA	ITALY	SOUTH AFRICA
BYELORUSSIAN SOVIET	JAMAICA	SPAIN
SOCIALIST REPUBLIC	JAPAN	SRI LANKA
CAMEROON	JORDAN	SUDAN
CANADA	KENYA	SWEDEN
CHILE	KOREA, REPUBLIC OF	SWITZERLAND
CHINA	KUWAIT	SYRIAN ARAB REPUBLIC
COLOMBIA	LEBANON	THAILAND
COSTA RICA	LIBERIA	TUNISIA
COTE D'IVOIRE	LIBYAN ARAB JAMAHIRIYA	TURKEY
CUBA	LIECHTENSTEIN	UGANDA
CYPRUS	LUXEMBOURG	UKRAINIAN SOVIET SOCIALIST
CZECHOSLOVAKIA	MADAGASCAR	REPUBLIC
DEMOCRATIC KAMPUCHEA	MALAYSIA	UNION OF SOVIET SOCIALIST
DEMOCRATIC PEOPLE'S	MALI	REPUBLICS
REPUBLIC OF KOREA	MAURITIUS	UNITED ARAB EMIRATES
DENMARK	MEXICO	UNITED KINGDOM OF GREAT
DOMINICAN REPUBLIC	MONACO	BRITAIN AND NORTHERN
ECUADOR	MONGOLIA	IRELAND
EGYPT	MOROCCO	UNITED REPUBLIC OF
EL SALVADOR	NAMIBIA	TANZANIA
ETHIOPIA	NETHERLANDS	UNITED STATES OF AMERICA
FINLAND	NEW ZEALAND	URUGUAY
FRANCE	NICARAGUA	VENEZUELA
GABON	NIGER	VIET NAM
GERMAN DEMOCRATIC REPUBLIC	NIGERIA	YUGOSLAVIA
GERMANY, FEDERAL REPUBLIC OF	NORWAY	ZAIRE
GHANA	PAKISTAN	ZAMBIA
GREECE	PANAMA	

The Agency's Statute was approved on 23 October 1956 by the Conference on the Statute of the IAEA held at United Nations Headquarters, New York; it entered into force on 29 July 1957. The Headquarters of the Agency are situated in Vienna. Its principal objective is "to accelerate and enlarge the contribution of atomic energy to peace, health and prosperity throughout the world".

Printed by the IAEA in Austria
September 1986

TECHNICAL REPORTS SERIES No. 267

METHODOLOGY AND TECHNOLOGY OF DECOMMISSIONING NUCLEAR FACILITIES

REPORT OF A TECHNICAL COMMITTEE MEETING
ON THE METHODOLOGY AND TECHNOLOGY
OF DECOMMISSIONING NUCLEAR FACILITIES
ORGANIZED BY THE
INTERNATIONAL ATOMIC ENERGY AGENCY
AND HELD IN VIENNA, 22–26 APRIL 1985

INTERNATIONAL ATOMIC ENERGY AGENCY
VIENNA, 1986

METHODOLOGY AND TECHNOLOGY
OF DECOMMISSIONING NUCLEAR FACILITIES
IAEA, VIENNA, 1986
STI/DOC/10/267
ISBN 92-0-125286-2

FOREWORD

The decommissioning and decontamination of nuclear facilities is a topic of great interest to many Member States of the International Atomic Energy Agency (IAEA) because of the large number of older nuclear facilities which are or soon will be retired from service. In response to increased international interest in decommissioning and to the needs of Member States, the IAEA's activities in this area have increased during the past few years and will be enhanced considerably in the future. A long range programme using an integrated systems approach covering all the technical, regulatory and safety steps associated with the decommissioning of nuclear facilities is being developed. The database resulting from this work is required so that Member States can decommission their nuclear facilities in a safe time and cost effective manner and the IAEA can effectively respond to requests for assistance.

A Technical Committee Meeting on the Methodology and Technology of Decommissioning Nuclear Facilities was held in April 1985 at the Agency's Headquarters in Vienna, Austria. The meeting was attended by 28 experts from 13 Member States and one international organization. The participants discussed and redrafted a preliminary report on the subject by the consultants P. De (Canada), W. Diefenbacher (Federal Republic of Germany), T.S. La Guardia (USA) and J. Griffin (UK) and the IAEA Scientific Secretary M.A. Feraday. After the meeting the report was revised by the IAEA Secretariat and the final report was approved by all participants.

The report is a review of the current state of the art of the methodology and technology of decommissioning nuclear facilities including remote systems technology. This is the first report in the IAEA's expanded programme and was of benefit in outlining future activities. Certain aspects of the work reviewed in this report, such as the recycling of radioactive materials from decommissioning, will be examined in depth in future reports.

The information presented should be useful to those responsible for or interested in planning or implementing the decommissioning of nuclear facilities.

CONTENTS

1. INTRODUCTION

The decommissioning of nuclear facilities is a topic of great interest to many Member States because of the large number of facilities which have been built and will eventually have to be retired from service. The term 'decommissioning', as used in the nuclear industry, means the actions taken at the end of a facility's useful life to retire the facility from service in a manner that provides adequate protection for the health and safety of the decommissioning workers, the general public, and for the environment. These actions can range from merely closing down the facility and a minimal removal of radioactive material coupled with continuing maintenance and surveillance, to a complete removal of residual radioactivity in excess of levels acceptable for unrestricted use of the facility and its site.

Since the International Atomic Energy Agency first included the subject of decommissioning in its programmes in 1973 nine decommissioning documents [1–9] related to the IAEA's programme have been published. These reports summarize the work done by various Technical Committees, Advisory Groups of Experts and International Symposia, and also give definitions and guidance for the subsequent activity of the Agency in this field [2, 9].

Over a hundred nuclear facilities have been or are being decommissioned (see Annex A). Although no large power reactor has been completely dismantled, technical experts agree that sufficient experience has been gained so far to demonstrate that such dismantlement can be carried out without unacceptable impact on man or his environment. Conceptual studies and projects support this viewpoint [10–13]. Even though progress has been made in the development of the technology and methodology of decommissioning, further work is required to improve equipment and techniques, reduce costs and exposures and gain experience with larger facilities.

IAEA Safety Series No.52 [7] provides an outline of the principles and factors to be considered in the decommissioning of land based nuclear reactors in a safe and orderly manner. The report contains general discussions on the planning (including facilitation of decommissioning at the design step), management, radiation and quality assurance and release criteria needed to carry out a decommissioning project successfully.

Technical Reports Series No. 230 [9] provides an information base on the technical considerations important to decommissioning, methods available for decontamination and disassembly of a nuclear facility, the management of the resulting radioactive wastes, and areas of decommissioning methodology where improvements can be made.

Another Technical Reports Series [14] deals more specifically with decontamination techniques and the management of waste arising from these activities.

The purpose of the present report is to complement and reinforce the data published in previous reports and, in particular, to provide a database for those areas of decommissioning technology where improvements might be made or where more detailed information is required to assess the scope of future work. Decommissioning areas that are well covered elsewhere, for example decontamination [8–15], will also be briefly summarized in this report in line with the IAEA's integrated systems approach being used to develop the database required for the wide ranging topics associated with decommissioning.

2. SCOPE

This report is a review of the current state of the art of the methodology and technology of decommissioning nuclear facilities including such advanced technologies as robotics and remote systems technology applied to decommissioning. The information presented should be useful to those responsible for, or interested in, planning or implementing the decommissioning of all types of facilities with the exception of mines. However, it should be used in conjunction with other published technical information and past experience gained with nuclear facilities, especially plants having similar characteristics or designs.

3. DEFINITIONS OF THE BASIC STAGES OF DECOMMISSIONING

The term 'stage' as used herein implies a set of conditions at the plant and does not necessarily imply a continuing stepwise procedure. The three stages as described in Ref. [9] are set out below.

When a plant is retired from service the nuclear fuel or radioactive materials in the process systems as well as radioactive waste produced in normal operation should first be removed by routine operations. However, for plants that handled fissile material the possibility that some residual fissile material still remains must be given serious consideration during decommissioning and suitable precautions taken.

Each of the three decommissioning stages of a nuclear plant can be defined by two parameters as follows:

— The physical state of the plant and its equipment;
— The surveillance, inspections and tests necessitated by that state.

3.1. Stage 1 decommissioning

The first contamination barrier is kept as it was during operation, but the mechanical opening systems are permanently blocked and sealed (valves, plugs, etc.).

The containment building is kept in a state appropriate to the remaining hazard and the atmosphere inside the building is subject to appropriate control.

Access to the inside of the building is subject to monitoring and surveillance procedures.

The unit is under surveillance and the equipment necessary for monitoring radioactivity both inside and outside the plant is kept in good condition and used when necessary and in accordance with national legal requirements. Inspections are carried out to check that the plant remains in good condition. If necessary, checks are carried out to see that there are no leaks in the first contamination barrier and the containment building.

3.2. Stage 2 decommissioning

The first contamination barrier is reduced to minimum size and all parts easily dismantled are removed. The sealing of that barrier is reinforced by physical means and the biological shield in a reactor is extended if necessary so that it completely surrounds the barrier.

After decontamination to acceptable levels, the containment building and the nuclear ventilation system may be modified or removed if they are no longer required for radiological safety. Depending on the extent to which other equipment is removed or decontaminated, access to the former containment building, if left standing, can be permitted.

The non-radioactive buildings or equipment in the plant may be converted for new purposes.

Surveillance around the barrier can be relaxed but it is desirable for periodic spot checks to be continued as appropriate, together with surveillance of the environment. External inspection of the sealed parts should also be performed.

3.3. Stage 3 decommissioning

All materials, equipment and parts of the plant in which activity remains significant despite decontamination are removed. In all remaining parts contamination has been reduced to acceptable levels.

The plant and site are released for unrestricted use. From the point of view of radiological protection, no further surveillance, inspection or tests are necessary.

In some cases the whole plant, including inactive components, may be dismantled to make room for a replacement facility or other usage.

4. CONSIDERATIONS AFFECTING DECOMMISSIONING CHOICES

The selection of the decommissioning process for a nuclear facility should be based on the results of a comprehensive comparison among various alternatives. This comparison should be made in the light of the national policy, which is the responsibility of the government and takes into account public opinion, and by analysing:

(a) the physical and radiological conditions of the facility after final shutdown and their subsequent evolution;
(b) the nuclear and industrial safety requirements including the results of a corresponding risk analysis;
(c) the issues connected with waste management (treatment conditioning, transport, intermediate storage, disposal);
(d) the possibility of reusing components, tools, buildings and land and of recycling materials;
(e) the availability of qualified and experienced personnel and suitable technical means;
(f) the cost estimates, the availability of funds and the compatibility with the financial management of the organization;
(g) the social and environmental impacts of the proposed activities.

Among the factors which have a bearing on the above comparison, the following are more extensively discussed in this report:

— National policy for decommissioning (Section 5.1)
— Estimation and verification of radioactive inventory (Section 5.2)
— Criteria for release of equipment, materials and sites (Section 5.3)
— Characterization of the radioactive inventory in decommissioning wastes (Section 5.4)
— Waste management (Section 5.5)
— Availability of suitable decontamination methods (Section 5.6) and disassembly techniques (Section 5.7)
— The long term integrity of buildings, structures and materials (Section 5.8)
— Decommissioning cost estimating and financing (Section 5.9).

Other important factors including items such as health and safety, planned use of the site, environmental and social considerations, and optimization are discussed briefly below.

(a) Health and safety

The primary concern in any decommissioning programme is to protect the health and safety of the workers and the public. Occupational exposures resulting from decommissioning activities can be quantified for each alternative. Experience during the replacement of major plant components at different types of facilities has shown that both direct and airborne (inhalation) radiation doses to the workers can be controlled safely. Advanced dismantling techniques such as remote tooling, shielding and preplanned exposure reduction procedures are some of the

4

methods that can be implemented to keep exposures 'as low as reasonably achievable (ALARA)'. This experience has also shown public exposures to be minimal during the decommissioning process.

The tools and techniques for decommissioning are available to remove radioactivity safely so as to permit unrestricted access to the facility. Consequently, post-decommissioning public safety can be achieved with present day techniques. The integrated costs and benefits to achieve these objectives are part of the overall decommissioning selection process.

(b) Owner's planned use of the site

Nuclear sites represent a valuable resource since they usually are in areas of low seismic activity, close to cooling water and have access to major freight transport routes, etc. Owners of nuclear facilities recognize the value of placing replacement power reactors or process facilities on the same location in view of public acceptance of the site. If available land is limited, the owner may decide to completely dismantle the plant, including inactive components, to make room for a replacement facility.

If the nuclear facility is adjacent to other operating facilities that will remain in service, continuing maintenance, surveillance and security for Stage 1 or 2 decommissioning options may be provided by personnel at the operating facility at a relatively small incremental cost.

The shut down nuclear facility may also have useful items such as turbines, water treatment systems, warehouses, administration buildings, or other structures suitable for conversion to non-nuclear uses. In that case the owner may decide not to dismantle these buildings.

(c) Environmental and social considerations

Environmental assessments will have to be carried out for the possible decommissioning alternatives. However, there will be significant differences among these alternatives, depending on site, type of plant and operational history.

Issues such as potential increase in land usage, aesthetic impact of a shut down facility, public acceptance and waste transport routes in congested city areas must be evaluated in making the choice of alternatives.

(d) Optimization

A decommissioning plan, which is prepared to implement a programme, must take into consideration the foregoing factors to satisfy regulatory requirements, public health and safety, and costs. Proper planning can optimize these factors to develop a responsive decommissioning plan in a cost effective manner. If several decommissioning alternatives are possible, the owner may opt for combinations of alternatives to suit his needs best as influenced by requirements for specialized dismantling equipment, occupational exposures, surveillance costs, interest rates, taxes, etc.

5. METHODOLOGY AND TECHNOLOGY OF DECOMMISSIONING

5.1. National policy for decommissioning

The development of a safe, efficient, viable and cost effective strategy for the decommissioning of no longer needed nuclear facilities in a country can best be done within the overall framework of an integrated national nuclear policy. The major objective of this policy should be to ensure that all the important facets are developed on a co-ordinated time-scale and in such a manner as to meet the national requirements.

The IAEA is available to give further guidance for the safe decommissioning of all nuclear facilities including those outside the nuclear fuel cycle. Decommissioning of nuclear facilities used in medicine, industry and research is also of interest in developing countries.

Although the components of a national policy for decommissioning may vary from country to country depending on national conditions, the following should in general be provided:

(a) A framework of laws and regulations within which a comprehensive decommissioning programme could develop. This framework should also clearly define the division of responsibilities between the policy makers, regulators and those doing the decommissioning. In addition, matters of national policy such as the optimization of exposure levels, de minimis criteria, the number, type and location of disposal facilities must be developed within this framework as well as the means of ensuring financial responsibility for the final decommissioning of individual facilities.

(b) A rational set of limits and criteria regarding allowable levels of residual radioactive substances in facilities and on materials, equipment and sites that are to be released for unrestricted use.

(c) Facilities to dispose of the wastes arising from decommissioning activities safely.

(d) Requirements for the organization carrying out decommissioning activities to demonstrate its technical and financial capacity to accomplish such activities.

(e) Suitable quality assurance and quality control requirements.

For each nuclear facility information about construction and operation should be carefully collected and organized so that it can be recovered later. This will be essential to provide a sound decommissioning plan, especially if decommissioning is delayed for long periods. The establishment of a suitable integrated databank on the national level should also be considered in Member States that have many facilities to be decommissioned.

In the following sections of this report some important factors in a decommissioning project are examined in more detail, taking into consideration past and present developments in Member States. However, those items that have been well covered elsewhere will only be briefly discussed.

5.2. Estimation and verification of radioactive inventory

The radioactive inventory in nuclear facilities to be decommissioned can be divided into two categories: (1) radioactivity induced by neutron activation of certain elements in reactor components and adjacent structures and (2) the radioactive substances deposited on the internal and external surfaces of various systems as contamination. Included in this category are daughter radionuclides which become significant after periods of decay.

As indicated in Section 3, it is assumed that the nuclear fuel and process fluids have been removed from the plant before the start of decommissioning activities; however, in some cases residues of these materials may remain which should be included in the radioactive inventory.

A good estimate of the amount and type of radioactivity in a nuclear facility is important because it can directly affect the whole approach to decommissioning including the choice of the time to start decommissioning and the desirability of delay between stages. In addition, such an estimate will be a great asset in the planning execution to ensure that the facility be decommissioned in a safe, economic and timely manner. This information will assist the planners in determining factors such as the need for decontamination, shielding or remotely operated equipment, shipping and disposal, and potential radiation exposures to the work force. These and other factors are discussed in the following sections.

The facility inventory should include detailed inventories for individual components and should describe specific radionuclide content, chemical form, physical form and volume distribution. The inventory should be maintained current as decommissioning progresses, and in cases of Stage 1 or 2 decommissioning the inventory must be projected into the future to demonstrate the required period of confinement of the radionuclides. More detailed information regarding distribution of radionuclides is required for Stage 3 decommissioning.

If the residual radioactivity consists mainly of short lived radionuclides, a significant reduction in radioactive inventory can be achieved by first decommissioning to Stage 1 to allow for natural decay. This will reduce occupational exposure to the workers, the amount of radioactivity contained in the waste, and the amount of material requiring disposal as radioactive waste.

TABLE I. MAJOR RADIONUCLIDES WHICH COULD BE PRESENT DURING THE DECOMMISSIONING OF NUCLEAR FACILITIES

Isotope	Half-life (years)	Principal decay modes	Associated γ energy (MeV)	Material (plants) where isotopes may occur
(A) Neutron activation products found in nuclear reactors[a]				
H-3[b]	12.3	β^-	–	C, O, S[b]
C-14	5730	β^-	–	G, M, S
Na-22[b]	2.6	β^+, EC	0.51, 1.28	O
Cl-36	3.1×10^5	β^-, EC	–	C
Ar-39[b]	269	β^-	–	C
Ca-41	1×10^5	EC	–	C
Ca-45	0.4	β^-	–	C
V-49[b]	0.9	EC	–	S[b]
Mn-54	0.9	EC, γ	0.83	A, M, S
Fe-55[b]	2.7	EC	–	C, M, O, S[b]
Co-57[b]	0.7	EC, γ	0.12, 0.14	S[b]
Co-60[b]	5.3	β^-, γ	1.2, 1.3	C, M, O, S, Z
Ni-59	7.5×10^4	EC	–	C, M, O, S, Z
Ni-63[b]	100	β^-	–	C, M, O, S[b]
Zn-65	0.7	β^+, EC, γ	0.51, 1.12	A
Zr-93	1.5×10^6	β^-	–	O, Z
Nb-94	2×10^4	β^-, γ	0.70, 0.87	M, O, S, Z
Mo-93	3.5×10^3	EC, γ	0.3	M
Ag-108m	130	EC, γ	0.4, 0.6, 0.7	M, O, S
Ag-110m	0.7	β^-, γ	0.6, 0.9	M, O, S
Ba-133[b]	10.7	EC, γ	0.08, 0.36	C
Sm-151	93	β^-, γ	0.02	C
Eu-152	13.4	EC, β^-, γ	0.1	C, G
Eu-154	8.2	β^-, γ	0.1, 1.3	C, G
(B) Uranium & transuranic elements found in facilities handling fissile materials				
				Plant type[c]
U-232	72	α, γ	0.06, 1.3	1
U-233	1.6×10^5	α, γ	0.04, 1.0	1
U-234	2.4×10^5	α, γ	0.05, 0.1	1
U-235	7×10^8	α, γ	0.2	3
U-237	0.02	β, γ	0.2	2
U-238	4×10^7	α, γ	0.05	2, 3
Np-237	2.1×10^6	α, γ	0.02, 0.08	2, 3
Pu-238	87.7	α, γ	0.04, 1.1	2, 3
Pu-239	2.4×10^4	α, γ	0.05	2, 3
Pu-240	6537	α, γ	0.04, 0.9	2, 3
Pu-241	14.7	α, β^-, γ	0.03, 0.15	2, 3
Pu-242	3.8×10^5	α, γ	0.04	2, 3
Am-241	432	α, γ	0.05, 0.8	2, 3
Am-243	7380	α, γ	0.07, 0.6	2, 3

TABLE I (cont.)

(C) Fission products found in reprocessing plants

In addition to uranium and transuranics shown in Section (B), the following fission products and their daughters could be of concern during decommissioning of reprocessing plants

Sr-90 $(29 \text{ a}, \beta^-)$ Cs-137 $(30 \text{ a}, 0.7 \text{ MeV } \beta^-, \gamma)$

Ru-106 $(1 \text{ a}, \beta^-)$ Ce-144 $(0.7 \text{ a}, \beta^-, 0.1 \text{ MeV } \gamma)$

Cs-134 $(2 \text{ a}, 0.8 \text{ MeV } \gamma)$

---

[a] Fission products are also often found in nuclear reactors as a result of defects in the fuel cladding.

[b] Important in fusion plants also.

[c] Plant types: 1 = Plant handling Th-U-233 fuel
                       2 = Plant handling Pu-U fuel
                       3 = Plant handling enriched U fuel

C = Concrete, G = Graphite, O = Other, A = Aluminium, Z = Zr alloys, S = Stainless steel, M = Mild steel, EC = Electron capture.

On the other hand, if the residual activity consists mainly of longer lived radionuclides, particularly transuranic elements, a significant reduction in inventory would not be achieved by delaying the completion of decommissioning. Furthermore, the containment integrity of the facility would probably not be adequate to contain these longer lived radionuclides for periods adequate to get significant natural decay.

It should be noted that cost savings resulting from delaying the completion of decommissioning will be offset somewhat by costs associated with the surveillance and maintenance required to prevent the facility from deteriorating.

5.2.1. Sources of radioactive inventory

Some nuclides contained in materials exposed to the neutron flux in nuclear reactors, accelerators or fusion devices are transformed into radioactive isotopes. The level and type of radionuclides found in neutron irradiated material depend on: (1) the nuclides in the material, (2) the duration of the exposure and the intensity of the neutron flux and, (3) the energies of the incident neutrons. The relevant radionuclides composing the induced radioactivity in reactor and accelerator components and adjacent structures are shown in Table I. In other types of facilities, activation products are not a serious concern during decommissioning.

Radioactive contamination of internal surfaces of reactor systems is caused by the deposition from the reactor coolant of neutron activated particles and dissolved elements, and from fission products and actinides released when there is a failure of the fuel cladding. External surface contamination in nuclear plants is primarily due to leakage and spills from reactor systems, and to maintenance and waste management activities.

Radioactive contamination in other fuel cycle facilities such as fuel fabrication and reprocessing facilities, glove box lines, and storage basins is due to deposition of radioactive materials from the process stream. This contamination generally consists of uranium, thorium and plutonium and their daughters in conversion, enrichment and fabrication plants. Fission product contamination is also present in reprocessing plants.

In facilities not associated with the nuclear fuel cycle the radioactivity arises from radionuclides in the process stream, for example ^{99}Mo in a radiopharmaceutical hot cell facility.

5.2.2. Application of inventory data

The data determined from an accurate assessment of the radioactive inventory are important for a variety of reasons. These data will play a major role in deciding on the overall decommissioning approach and determining whether or not Stage 3 should be delayed. The inventory data will also provide the information needed to plan decommissioning activities, including the scheduling and manpower requirements, particularly with respect to exposures in the most radioactive areas.

For personnel exposures the data are necessary to prepare procedures for keeping exposures as low as reasonably achievable (ALARA). Data on the decay of radiation fields with time will indicate how long a delay is required before simplified or manual removal or dismantling can be initiated. Alternatively, the data could be used to assess the need for supplementary shielding of piping, components, etc., during dismantling.

The data provide a basis for determining the amount of decontamination to be performed and the method of decontamination to be employed. For example, on the basis of the data, a simple water flush of a piping system before dismantling may be shown to be more cost effective than harsher methods.

The type and concentration of radionuclides in contaminated or activated material from a nuclear facility will have a direct bearing on the method of packaging and shielding for shipment as well as for disposal. An accurate inventory of radionuclides is necessary to demonstrate compliance with regulations governing the disposal of radioactive materials. The inventory may, in some cases, show that burial of certain material in a radioactive waste repository is not necessary because of low radioactivity levels. In such cases substantial financial savings may be realized.

The inventory estimate is also needed to assess the radiation exposure of the general public due to the transport and disposal of waste and due to expected and accidental releases of liquid and gaseous waste associated with decommissioning activities.

5.2.3. Radionuclides of concern

5.2.3.1. Nuclear reactors

The radionuclides of concern will vary considerably depending upon the type of facility being decommissioned. In the case of a nuclear reactor the major radionuclide of concern during the first fifty to one hundred years is ^{60}Co, which emits high energy gamma rays (Table I). Thereafter other radionuclides such as ^{63}Ni and ^{108}Ag^m become predominant.

The dose levels of gamma radiation from activation products such as ^{60}Co in the reactor vessel and internal components of the reactor vessel will determine the amount of remote operation and shielding that is required. In the longer term other radionuclides such as ^{63}Ni, ^{59}Ni, ^{108}Ag^m and ^{14}C are of greater concern because their long half-lives will render them radioactive for hundreds of years.

Nickel-59 has not been of major importance previously because reactors had not been in operation long enough to create significant quantities of this radionuclide. However, a reactor in operation for 40 years may contain significant amounts of ^{59}Ni in certain reactor vessel components. Carbon-14 is of concern for graphite moderated reactors. Other radionuclides encountered in neutron irradiated material are shown in Table I.

Of importance, but not well documented, is the radiation resulting from the activation of trace elements to produce radionuclides such as ^{94}Nb that could give rise to significant activity in the very long term. Trace elements may be natural impurities or may be introduced from scrap metal added to virgin metal during manufacture. For example, niobium is added during the manufacture of steel to improve welding characteristics. When scrap steel is added to the virgin metal melt, diluted niobium remains in trace quantities [16].

Fission products and actinides may also be present in reactor facilities as a result of fuel failures. If the facility has experienced an accident during its operating lifetime, significantly higher inventories of fission products and actinides may be present.

5.2.3.2. Other nuclear facilities

In other nuclear fuel cycle facilities the radionuclides of concern will depend on which part of the cycle the facility was used for.

In the case of facilities at the front end of the nuclear fuel cycle, the members of the uranium and thorium decay chains will be the most important radionuclides (Fig. 1).

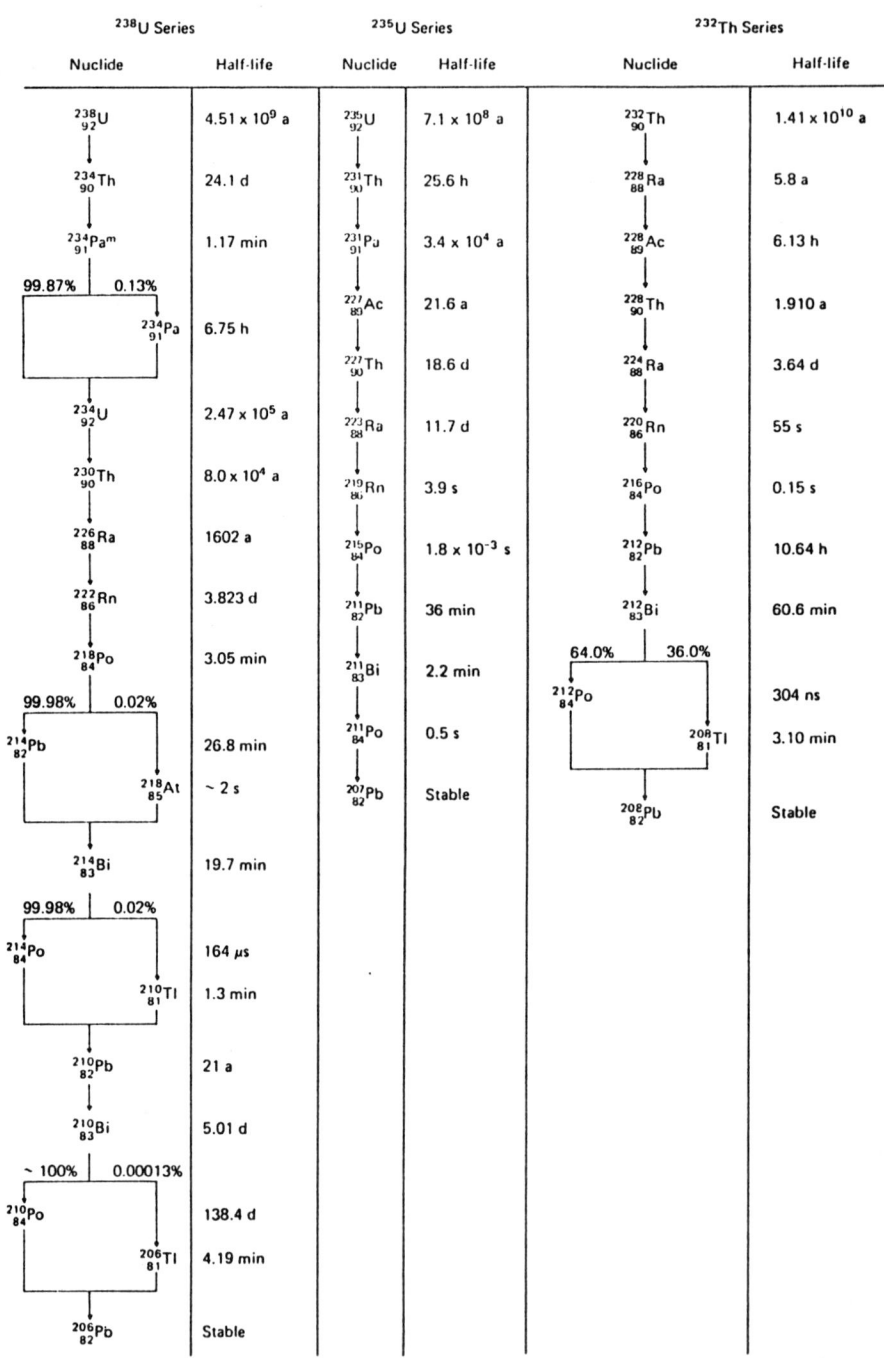

FIG. 1. Radioactive decay series for uranium and thorium [17].

For plutonium facilities or those handling highly enriched uranium, the actinides such as [239]Pu (Table I) will be the main problem. These radionuclides are much more difficult to handle than either fission products or activation products owing to their high specific radiotoxicity and the associated inhalation hazard. However, in reprocessing plants the fission and activation products also contribute significantly to the total potential hazard. For [233]U facilities the actinides and the decay product [208]Tl (2.6 MeV gamma energy) are the major concerns.

In the case of non-nuclear fuel cycle facilities the radionuclides of concern will depend on the product being used or produced. Examples of nuclides of importance in some plants include: [60]Co for gamma irradiation, ^{131}I and ^{51}Cr in radiopharmaceutical facilities and [226]Ra in gypsum residues from the phosphate industry.

In fusion facilities the major radionuclides of concern are activation products similar to those that occur in nuclear reactors.

5.2.4. Estimation of radioactive inventory

5.2.4.1. Activation products

In general, the radionuclide inventory in nuclear reactors is obtained by calculating the concentration of the radionuclides formed by neutron activation during operation. Such calculations require data on the size, weight and composition of the irradiated components, including data on trace elements and impurities and on the operating history of the reactor and the integrated neutron flux on the components. These calculations are usually performed on a computer using codes that have the capability to solve complex multiple energy group equations. Hand calculations can be used to provide conservative estimates.

The determination of elements present in the components is most accurately done by sampling at the time of decommissioning, by analysis of samples taken during construction or by analysis provided by the supplier [18–20]. If this information is not available, then compositions can be estimated from material specifications. Specific nuclides may be chosen for analysis because of their half-life, cross-section or abundance. In the absence of analysis, trace elements in material may be conservatively estimated by utilizing techniques which maximize individual elements [21].

The material densities (isotopic compositions) in the homogenized, exposed core zone and activated components may be calculated using computer codes such as the ORIGEN [22] code. The nuclear cross-section data for ORIGEN may be calculated from the ENDF/B [23] code. These data are then formatted into multigroup (typically 30 energy groups) codes such as ETOG [24].

The distribution of neutron flux within the core and in surrounding activated components can be calculated using computer codes such as ANISN [25].

TABLE II. ESTIMATED RADIONUCLIDE COMPOSITION ON THE
INTERNAL SURFACES OF A REFERENCE BWR AT FINAL SHUTDOWN
Based on sludge sample analysis [26]

Radionuclide	Shutdown	Fractional radioactivity at decay times of[a]			
		10 years	30 years	50 years	100 years
Cr-51	2.1×10^{-2}	–	–	–	–
Mn-54	3.9×10^{-1}	1.3×10^{-4}	–	–	–
Fe-59	2.5×10^{-2}	–	–	–	–
Co-58	9.3×10^{-3}	–	–	–	–
Co-60	4.7×10^{-1}	1.3×10^{-1}	9.1×10^{-3}	6.6×10^{-4}	9.2×10^{-7}
Zn-65	6.1×10^{-3}	1.5×10^{-7}	–	–	–
Zr-95	4.0×10^{-3}	–	–	–	–
Nb-95	4.0×10^{-3}	–	–	–	–
Ru-103	2.3×10^{-3}	–	–	–	–
Ru-106	2.8×10^{-3}	2.7×10^{-6}	–	–	–
Cs-137	3.4×10^{-2}	2.7×10^{-2}	1.7×10^{-2}	1.1×10^{-2}	3.4×10^{-3}
Ce-141	3.0×10^{-3}	–	–	–	–
Ce-144	$8.1 \quad 10^{-3}$	8.8×10^{-7}	–	–	–
Totals	1.0	1.6×10^{-1}	2.6×10^{-2}	1.2×10^{-2}	3.4×10^{-3}

[a] Dashes indicate values below 10^{-10}.

The neutron flux values at the components of interest are used in the ORIGEN
production–depletion calculations to obtain the radionuclide inventories in these
various reactor components.

Results of such analyses for 1100 MW(e) reactors operated for 30 effective
full power years show that PWR and BWR vessels and internals would contain
approximately 2×10^{17} Bq immediately after shutdown, not including the fuel.

Estimates of exposure rates from neutron activated components can be
made using established methodology and equations such as those presented in the
US DOE Decommissioning Handbook [10]. For neutron activated materials
the method consists of determining the volumetric thick slab source geometry
for components such as vessels and internals and calculating the buildup gamma
flux. Other components may dictate the use of point, line, planar or cylindrical
geometries as appropriate. Verification of the neutron activation may then be
made by direct measurement of surface exposure rates.

14

TABLE III. ESTIMATED RADIOACTIVITY DEPOSITION IN BWR PIPING CALCULATED FROM MEASURED CONTACT DOSE RATES ON THE OUTSIDE OF THE PIPE [26]

Pipe category	Nominal outside diameter (mm)	Wall thickness (mm)	Measured contact dose rate (mR/h)	Estimated deposition level (GBq/m^2)
Reactor water	610	59.5	700	41
Steam/air	914	20.4	70	0.2
Condensate	610	46.0	50	1.9

5.2.4.2. Internal surface contamination

Internal contamination levels on piping and components can be estimated from internal or external radiation measurements using simplified calculated correlations, or from samples taken from the pipe. The major contributor to the external exposure rate on reactor piping is from ^{60}Co and, accordingly, the correlation factor is strongly influenced by this radionuclide. The estimated contamination inventory on internal pipe surfaces of a BWR based on a sludge sample analysis is presented in NUREG/CR-0672 [26] and is shown in Table II.

Computer codes such as ISOSHLD [27] can be used to correlate external measured exposure rates to internal surface contamination. With appropriate assumptions regarding pipe and equipment geometry, the estimated results for a BWR are shown in Table III (from Ref. [26]).

Results such as these were used to estimate the total contamination in a 1155 MW(e) BWR to be 233 TBq (6300 Ci) in piping and equipment (excluding the reactor vessel and internals).

5.2.4.3. Estimation of fissile materials

In the decommissioning of reprocessing plants and other facilities which had previously processed highly enriched fissile material, some residual fissile material could remain. Therefore, during the decontamination of these facilities normal criticality procedures should be exercised and the procedures should not be carried out without expert advice.

5.2.5. Verification of radioactive inventory

The calculated estimates of radionuclide inventory should be verified wherever possible by means of direct measurement. After final facility shutdown a rigorous sampling and direct measurement programme should be carried out to verify that inventory by comparing direct measurements with calculated exposure rates [28–30]. Occupational exposure considerations should be taken into account when planning for direct measurements.

The verification measurements may be performed using a gamma ray spectrometer coupled to a multichannel analyser. For reactor vessels and internals the dose rates would generally prohibit direct personnel access. One method for measurement is to place a detector at the inside of a thin walled stainless steel tube that is sealed at the bottom. The bottom can be filled with lead shot to shield the detector from some of the back or bottom scattered radiation. The tube may then be inserted into the water filled (for personnel shielding) vessel to read the contact dose rates. The resulting gamma exposure rate measurements may then be used to verify or normalize the calculated material activation. If there are inconsistencies between measurements and calculated results, re-examination of the assumptions or model should be made.

5.2.6. Other hazardous materials

Before decommissioning, an inventory of other hazardous materials such as beryllium, asbestos, pyrophorics, flammables, and toxic chemicals should also be completed. Methods and costs of decommissioning as well as personnel safety could be significantly affected by the presence of such materials. These materials may or may not be radioactive as well.

5.3. Criteria for release of equipment, materials and sites

Development of criteria for the unrestricted release of materials, equipment and sites for reuse, recycle or disposal has been recommended in several studies [9, 31]. These criteria are needed so that:

(a) very low level wastes can be disposed of as landfill, which may be much cheaper than even the least expensive form of shallow land burial in a radioactive material repository, and

(b) valuable metals from facilities being decommissioned can be recycled and expensive equipment can be reused.

These criteria should be based on internationally accepted basic environmental principles such as those established by the ICRP and the IAEA [32–35]. Several international organizations are working on the establishment of such limits.

In the past the IAEA has had an interest in the development of the concept of 'de minimis quantities' for waste disposal. Previous IAEA working groups looked at the considerations required to define de minimis quantities of waste for disposal of radioactive waste in the sea, by landfill or by incineration [36, 37].

16

In current work 'exemption from regulatory control' is being used in preference to 'de minimis' since the latter term has come to mean universally trivial irrespective of the circumstances of application. The conditions for exemption make it clear that each situation in which an exemption is considered has to be analysed separately to see if exemption rules are complied with. However, this does not exclude the possibility of general exemptions for certain well defined types of waste. In the current programme the philosophical basis for exemption rules is being developed and will be applied to practical waste management problems. A report is being prepared outlining the methods for deriving practical exempt quantities for application to terrestrial waste disposal of municipal landfill sites and for using incinerators to reduce the volume of the waste [38].

In 1986 an IAEA study will be initiated to apply the exemption rules and to develop modelling methods for evaluating practical exempt quantities for application to disposal of wastes from decommissioning and for the possible recycle of equipment and materials.

A working party sponsored by the Commission of the European Communities (CEC) is working on the establishment of radiological protection criteria for the recycling of materials from the dismantling of nuclear installations. The objective is to establish activity concentrations and surface limits that would be acceptable to all countries within the European Community and would allow recycle of valuable metals and transport of these materials across national borders for reuse without causing concern.

5.3.1. Current situation

The material arising from the decommissioning of a nuclear facility can be classified as **radioactive** (high, medium and low level) or as **inactive**. The inactive material can be: recycled, reused or disposed of as landfill.

Although all three methods have been used to manage inactive material from the dismantling of nuclear facilities (Section 5.5.3.3), this has only been done on a case by case basis using ad hoc release criteria. No country has yet established regulatory limits which would permit unrestricted release of all very low level radioactive waste.

In some countries criteria which have been used (or suggested for use) for the release of radioactive material on a case by case basis are generally in the range of:

— 0.37 to 3.7 Bq/g for specific activity,
— 0.37 to 3.7 Bq/cm² for beta/gamma surface contamination, and
— 0.037 to 0.37 Bq/cm² for alpha surface contamination,

where 3.7 Bq equals 100 pCi.

From a practical decommissioning viewpoint, in addition to the strictly environmental aspects, the following factors should also be considered in establishing criteria for unrestricted release:

- Availability of suitable field instruments which can measure radioactive levels as low as the selected limits ;
- Economic implications;
- Implications to radioactivity regulations used outside the nuclear industry.

5.4. Characterization of the radioactive inventory in decommissioning wastes

Estimation of the radiological inventory in a facility is a front end task required to define the operational decommissioning plan and estimate costs and radiological risks associated with the plan. Once the decommissioning programme is under way, regulatory, safety and waste disposal considerations require that wastes be monitored and characterized to a level not possible using only the original inventory estimates. The objective of this characterization is to ensure that the wastes will be handled and disposed of in a safe and economic manner. Therefore the characterization programme should be able to:

- segregate wastes into active and inactive (exempt from regulatory control) streams quickly and accurately so that inactive materials and equipment can be released for unrestricted use or for disposal in a sanitary landfill dump. Since the cost of waste disposal in a landfill site is usually smaller than the cost of even shallow land burial, the economic impact of segregating the large volumes of low activity wastes from decommissioning may be large. However, other costs, including those associated with monitoring and decontaminating the material to unrestricted release levels, should also be considered;

- determine the radiological characteristics of the decommissioning wastes which have been judged to be active. The authorities responsible for waste disposal sites will want to know both the major radionuclides present in the decommissioning wastes and the quantities of each radionuclide and the chemical form. These data are required so that the operators can ensure that the waste will go into the correct type of repository and that the total source term of radioactivity is known when the final safety assessment is done for the repository;

- characterize the site for final release.

5.4.1. Characterization methods

The methods and equipment used to characterize the radioactive wastes which arise from decommissioning will vary considerably, depending upon the type and complexity of the facility and the radionuclide mix in the plant.

The simplest type of decommissioning waste to characterize arises from facilities containing only one type of radionuclide, for example a ^{60}Co processing plant. Similarly, decommissioning wastes arising from plants which processed materials with a fixed ratio of alpha, beta and gamma activity would be relatively easy to characterize. Such plants could include: certain fuel fabrication facilities, mining/milling buildings and spent fuel storage facilities. In addition, pressure vessels and pressure tubes which have been decontaminated as well as concrete from biological shields should be amenable to this type of characterization since the ratio of radionuclides should not vary even though the concentrations may change. In all the above cases the quantity and quality of activity in unit packages of waste arising from the decommissioning of the facilities could be determined by gross gamma measurement and suitable modelling protocol.

The waste streams arising from the decommissioning of reactors, reprocessing plants and recycle fuel fabrication plants would be the most difficult to characterize, especially if the locations in the plant where the waste streams arise are not well defined. Complications arise because of the possible presence of streams having separated plutonium or variable mixes of alpha, beta and gamma activity.

In the following sections a review is made of the possible methods of measuring the activity in the waste, instruments currently available, means of establishing a waste characterization programme and relevant experience.

5.4.1.1. Gross gamma measurement

The easiest and least expensive means of characterizing radioactive waste is by measuring the gamma and X-ray intensity external to the package. This type of characterization requires that the waste arising from facilities has fixed ratios of alpha and beta to gamma activity which can be predetermined by laboratory measurements. However, in many cases these ratios are not constant (particularly for alpha radiation) and other methods will be required.

Measurements of total radiation fields due to gamma and X-radiation from waste containers can provide an acceptable estimate of the activity if the relationship between activity content and radiation field has been well established. Because the characteristics of the bulk of the waste are measured, this method is more suitable than sample spectrum analysis (Section 5.4.1.2) for the examination of heterogeneous low level wastes but is not as indicative of activity content as bulk spectrum analysis. As with bulk spectrum analysis the need for taking a sample is eliminated and the measurement time is short, so large volumes of waste can be processed relatively quickly.

A gross gamma reading alone will not indicate the nature and quantity of each of the major isotopes in a given waste package unless a previous detailed analysis of the waste stream was done to derive isotope concentrations comparable with total gamma readings. Such relationships lose their validity if the ratios of radionuclides in the waste deviate from those of the initial quantification analysis.

The accuracy of this method, then, is dependent upon factors such as container geometry, relative isotope mixture, distribution of activity in the package, background radiation and actual measurement procedure (probe distance, measurement points, instrumentation used, etc.). The successful use of total gamma measurement as a waste characterization tool requires that all of these factors remain constant or be adjusted for.

5.4.1.2. Gamma spectrum analysis

The most detailed analysis of the gamma emitting radionuclides can be obtained using gamma spectroscopy. This approach is required if the ratio of gamma emitters in the waste changes, for example if there are large variations in the ratio of ^{60}Co to ^{137}Cs.

In assaying wastes using gamma spectra it is neither necessary nor desirable to measure all the gamma emitting radionuclides and selection criteria must be established. In general, isotopes with half-lives of less than a year can be disregarded since they have little bearing on the potential detriment to man during most decommissioning operations and in waste disposal. The selection of the remaining radionuclides to be measured depends on the waste stream.

Accurate characterization requires that a small representative sample taken from the material stream or package be characterized. The spectrum of gamma radiation from the sample is measured and from it the constituents and their activities are deduced. Assuming that the sample is representative of the stream, the total activity per unit weight of stream can be calculated.

Such analysis, to be effective, generally requires the use of sophisticated equipment such as Ge detectors and multichannel analysers. The necessity of qualified personnel and a preferred laboratory environment makes this method time consuming and further increases the cost.

Although gamma spectroscopy gives an accurate determination of activity for gamma emitting isotopes, this method cannot detect the presence of non-gamma emitters. The overall accuracy of this technique, when applied under decommissioning conditions, is further limited by the small sample size analysed. The composition of low level wastes, which will constitute the major portion of total wastes, is known to vary considerably. Small sample size, therefore, unless supplemented by multiple sampling, will not necessarily give a true indication of the radionuclide content of a package or component.

Gamma spectrum analysis based on measurements of the external radiation from a waste package or component is also possible. This method is generally much less accurate than the specific sampling method, but does benefit from the fact that the entire bulk of the waste is assayed. Its use requires that final results be corrected for factors such as the effects of package geometry, package shielding, and self-shielding of wastes. The capital investment in energy sensitive, high resolution gamma detectors, multichannel analysers and other analytical control hardware remains the same as that for sample analysis.

To some extent, the combined advantages of the spectrum analysis and gross gamma measurement systems can be achieved through measurement with an energy sensitive detector and a single-channel analyser. Gamma rays of one or more energies are externally counted at various positions on the waste container. Given the proper pre-established relationships, the activity content can then be derived. For example, if ^{60}Co and ^{137}Cs are known to be the major isotopes present, two counts would be taken on the waste, one with the single-channel analyser set for 1.35 MeV (for ^{60}Co), the other with the analyser set for 0.6 to 0.7 MeV (for ^{137}Cs). This procedure has the advantage of bulk field count combined with the accuracy afforded by being able to count and integrate the photons from one or more specific isotopes in the waste. Measurement procedures and activity derivation factors must still be developed to account for the effects of package geometries, self-shielding and the presence of non-gamma emitting isotopes. The major advantage of this method over externally measured full spectrum analysis is in the lower cost of equipment.

The methods described lend themselves to high volume/production scale automation. Such adaptation, however, requires that all influencing factors such as package characteristics, presence of non-gamma emitters and self-shielding be accounted for in the control analysis logic.

5.4.2. Instrumentation

Accurate characterization of radioactive waste requires that the detector be suited to the energy levels of the radiation being emitted and that the resolution and accuracy of the detector be sufficient to meet the needs of the characterization programme.

A wide variety of instruments have been developed for measuring the radiation emitted from waste material. Three general categories of instruments have been used to measure this radioactivity; gass filled detectors, scintillation detectors or phosphors and solid state detectors. In general, the energy emitted during the interaction of the radiation with the material in the instrument is converted into an electrical pulse which can be recorded. The total radioactivity can be measured by summing the pulses over a fixed interval of time or by converting them to a pulse rate. In spectrum analyses the pulses are sorted out by energy level and the number of pulses at each level is stored separately using a pulse height analyser.

Some of the more common instruments that have been used to characterize waste during the decommissioning of facilities or during waste management operations at nuclear facilities or disposal sites are briefly described. These and other similar instruments have been used to measure the radiation arising from surfaces of equipment or facilities to determine whether or not such items could be released for unrestricted use.

5.4.2.1. Gas filled detectors

One widely used type of gas filled detector is the Geiger-Müller (GM) detector which consists of a sealed tube containing the counting gas, anode, cathode and a secondary gas to quench the discharge and prevent secondary discharges [39]. The counter is inexpensive, trouble free and generally used to measure gross gamma or beta/gamma. Certain designs can be used to detect alpha, beta and gamma radiation. The detectors can measure alpha above 3.5 MeV, beta above 35 keV and gamma above 6 keV.

Since GM detectors are incapable of resolution, they cannot be used for spectrum analysis. However, such instruments can be used effectively to measure the quantity of radiation in waste having fixed ratios of alpha and/or beta to gamma radiation, provided that:

(a) samples of the waste have been well characterized in the laboratory as to radionuclide mix and content
(b) the ratio of radionuclides in the stream does not change without one's knowledge
(c) the operators are well trained and use the instruments in a consistent manner
(d) the total weight of material is known.

The method is quite susceptible to error since it is subject to possible changes in radionuclide mix, at least in some plants, and to operator inconsistency.

Another type of gas filled detector which has been widely used is the gas flow proportional detector. This type of detector can be made with a thin window which increases the detection capability for alpha particles. A third type of gas filled detector, the ionization chamber, can be made in a variety of shapes and sizes, is extremely rugged and can operate over a wide range of temperature and gamma radiation levels.

5.4.2.2. Scintillation detectors

A zinc sulphide scintillator may be used during decommissioning [39] as an alpha survey meter. Plastic scintillators have been used as beta detectors and have a 100% efficiency for particles as low as 60 keV which pass through the protective foil window. The plastic is rugged and large detector sizes are possible; however, resolution for gamma rays is poorer than for NaI.

Sodium iodide is the best scintillation detector material for high energy gamma detection and overall gamma ray resolution. The best resolution achievable is about 7% (for 662 keV ^{137}Cs gamma). NaI detectors are widely used in gamma spectrometers where portability and counting efficiency are the key parameters.

FIG. 2.   A completely portable high resolution germanium diode spectrometer system capable
of measuring transuranics, activation products and fission products (including ^{90}Sr) at
sensitivities below the uncontrolled release criteria limits for the USA [40].

23

NaI detectors could best be used to segregate active/inactive waste since their detection ability and resolution are sufficient to define wastes above and below a de minimis radiation level in the range described in Section 5.3.1. Detailed characterization of active wastes is best left to the higher resolution solid state detectors. The reservation regarding alpha emitting waste as described in Section 5.4.1.1 applies to NaI detectors.

5.4.2.3. Solid state detectors

These detectors are essentially solid state ionization chambers, but with a higher efficiency owing to a solid, rather than a gas, detection medium. They are operated at liquid nitrogen temperatures to reduce thermal noise.

The decision on which type of solid state detector to adopt for a particular programme will depend on items such as which radioisotopes are to be quantified, which measurement method is to be implemented (gross gamma, gross alpha, single-channel analysis, multichannel analysis), throughput, degree of automation and cost. While the range and resolution of the instruments are prime considerations, the electronics (i.e. preamplifiers, amplifiers, analysers, etc.) required with each of the different types must also be carefully evaluated since they will affect both the cost and the flexibility of the final instrumentation set-up.

Two completely portable high resolution germanium diode spectrometer systems (Fig. 2) capable of measuring transuranics, activation products and fission products at very low levels have been developed for decommissioning activities [40]. Both units are capable of measuring gamma and X-rays. The data acquisition is accomplished using a 4096-channel analyser and a cassette tape recording for subsequent computer reduction and analysis.

5.4.2.4. Special alpha techniques

Wastes contaminated with transuranic (TRU) nuclides, for example ^{239}Pu, ^{241}Pu and ^{241}Am, are difficult to detect, especially in the presence of certain fission products. The selection criteria for the measurement of TRU are not as clear cut as for activation and fission products, especially if the waste is in packages. The only radiation that can be measured unequivocally from TRU wastes in the presence of other radionuclides which usually occur are the neutrons emitted spontaneously from the waste (passive) or those emitted following interrogation of the waste with pulsed neutron or photon sources (active).

Passive neutron techniques can be used to measure many of the TRU nuclides in low or high gamma activity wastes. Such techniques are being developed for the verification of the TRU content of waste packages (see Section 5.4.4).

5.4.3. Establishing a waste characterization and verification programme

The essential steps of a waste characterization and verification programme for a decommissioning project consist of:

(a) conducting extensive sampling and laboratory analyses
(b) dividing the intact facility into waste streams which are likely to have similar radionuclide mixes
(c) obtaining suitable criteria which would permit inactive waste to be segregated from active waste and permit the active waste to be classified according to disposal facility
(d) obtaining or developing instruments and equipment which would allow the waste to be segregated as in (c) as quickly, economically and accurately as possible, keeping in mind ALARA and other commitments of the decommissioning programme
(e) developing a well trained monitoring crew.

Item (c) is discussed briefly in Section 5.3. The monitoring personnel (e) should understand the equipment and criteria used for segregating the waste and should supervise and give direction to a team of operators who actually do the work.

The data required to divide the intact facility into discrete waste streams with similar radionuclide mixes could be obtained from the radiological inventory estimates and sampling procedures initially done in the facility to be decommissioned. The volumes of these streams can be calculated from the construction drawings. Depending on the type of facility, radionuclides present and the thoroughness of the initial survey, samples of the waste streams from the dismantling may or may not have to be analysed in detail to identify all the important radionuclides present using spectrographic and possibly chemical analyses. This type of confirmation would not be required for plants with fixed mixes of radionuclides, for example a natural uranium fabrication plant, but would be required for more complex plants such as reprocessing plants.

Once the streams have been defined, the type of instrumentation and sampling protocol required to verify the activity can be determined. Depending on the waste stream, monitors for gross gamma, beta and possibly alpha may be required as well as equipment for doing spectrum analyses on some or all of the waste. Depending on the form of the waste and/or the part of the plant or site being monitored, the physical design of the monitor could vary from hand held types to a more advanced production line monitoring assembly (see Section 5.4.4).

While for small samples the most accurate method of characterizing the radio-activity in waste is gamma spectrum analysis, this is not practical for large volumes of solid waste. For waste streams which require spectrum analysis, economics in time and labour can be achieved by combining several techniques. For example, in a specific waste stream having reasonably fixed mixes of gamma radionuclides, but variations in activity levels, the total activity content can be estimated fairly well using gross gamma measurements on most of the waste and spectrum analysis on selected samples from the wastes containing the highest activity levels.

Sources suspected of containing significant quantities of radioisotopes which are difficult to measure with gamma detection equipment, for example weak or non-gamma emitters, would require further analysis to develop effective scaling factors to relate gamma levels to the content of the other radionuclides. Further analysis, enhanced by chemical separation of the relevant isotopes and combined with an isotope behaviour analysis, would yield relatively constant relationships between the hard to detect radioisotopes and those that can be readily measured. Scaling factors can be established from these relationships. Subsequent surveys of wastes being removed from specific source areas or systems can then be accomplished with more practical gamma measurement techniques. The concentrations of detectable isotopes would be used, with the scaling factors, to derive the concentrations of non-gamma isotopes.

Once specific waste streams are characterized in this manner, regular assaying of the waste packages by external gamma measurement, either with a multichannel or single-channel analyser, will be sufficient to ensure compliance with safety and regulatory criteria. The initial source characterization would show which radioisotopes are present and, hence the specific gamma energy regions to be surveyed if single-channel analysis is used. Periodic laboratory analysis of wastes would still be required to confirm the validity of the scaling factors and the assumed radioisotope mix. The frequency of both the regular waste surveys and the periodic confirmatory analysis should be high during the initial periods of waste generation and can be permitted to drop off as confidence in the methods grows.

The degree of reliability of these mixed survey methods, and the resultant time and cost savings, depend primarily on the consistency of the waste streams. This requires that the waste be sorted into streams as homogeneous as possible. The ability to effectively define general sources in such a convenient manner will vary from facility to facility. Any difficulty in establishing a suitable definitive source base characterization can be compensated for by adopting more conservative scaling factors and/or more precise gamma measurement procedures. The latter applies to single-channel analysis and would be achieved by regularly surveying beyond the energy regions indicated by the detailed source base analyses. Such a measure would result in more operator time being required, while the use of conservative scaling factors would ultimately result in less efficient use of disposal space.

In the special case of fissile material facilities, the requirements for criticality control may necessitate measurement of all the waste streams from selected areas.

5.4.4. Relevant experience

Most waste arising from the decommissioning of nuclear facilities and the cleanup of sites in the past has been characterized for disposal using hand held instruments such as those described in Section 5.4.2 along with suitable sampling protocols. Similar instruments are also being used to characterize radioactive waste from nuclear power plants and other such facilities.

However, waste classification requirements imposed by recent legislation in certain countries [41] indicate that more detailed analyses of wastes going to disposal sites will be required to ensure that the concentrations of important radioisotopes are known within certain limits. The technology to accurately assay waste packages within the required confines is not currently available for large volumes of waste. However, attempts are in progress in various countries to develop the instrumentation and assay procedures required to verify the radionuclide content of low level waste from nuclear plants. This technology should be applicable to wastes arising from decommissioning.

In the USA, a high throughput demonstration unit for the verification of the radionuclide content of waste packages is being assembled [41]. This unit (Fig. 3) incorporates a turntable that rotates the waste package through 360° relative to a high resolution germanium diode gamma ray spectrometer while traversely scanning the package up and down. The unit is being evaluated for its capability to quantitatively measure ^{90}Sr and TRU content as well as activation and fission products. The passive neutron detection system would be used to quantitatively assay TRU (but not ^{235}U, ^{241}Am, ^{239}Pu or ^{241}Pu) concentrations. This particular system was developed for low levels of radioactivity and is calibrated with standards of ^{60}Co, ^{90}Sr, ^{134}Cs and ^{137}Cs at about 0.4 kBq/m³ or TRU at about 0.4 kBq/g.

At LASL/ORNL [42, 43] a neutron assay system to provide assays of the TRU and uranium content at the 3.7 kBq/g or less level in large crates has been built and tested for over a year. On average, each of the 3.5 m³ crates contained about 1000 kg of waste. The crates are moved into the assay chamber on an air pallet. A high sensitivity drum assay system incorporating pulsed thermal neutron interrogation and gamma ray spectroscopy has also been developed.

In the case of large quantities of fissile materials there are severe problems in using thermal neutron interrogation methods because of the large self-shielding effects of lumps of fissile material. Further development work is required for assay systems to cover this situation.

Since precision waste package assay instrumentation can most easily be developed for standard packages, consideration should be given to using two or three standard size and shape packages for most waste if possible. In the USA two package types (200 L drums and 1.2 X 1.2 X 2.4 m plywood boxes) are used for over 85% of the waste being received at low level non-TRU waste disposal sites. Development of the two high throughput assay systems described above is being standardized to these sizes of packages.

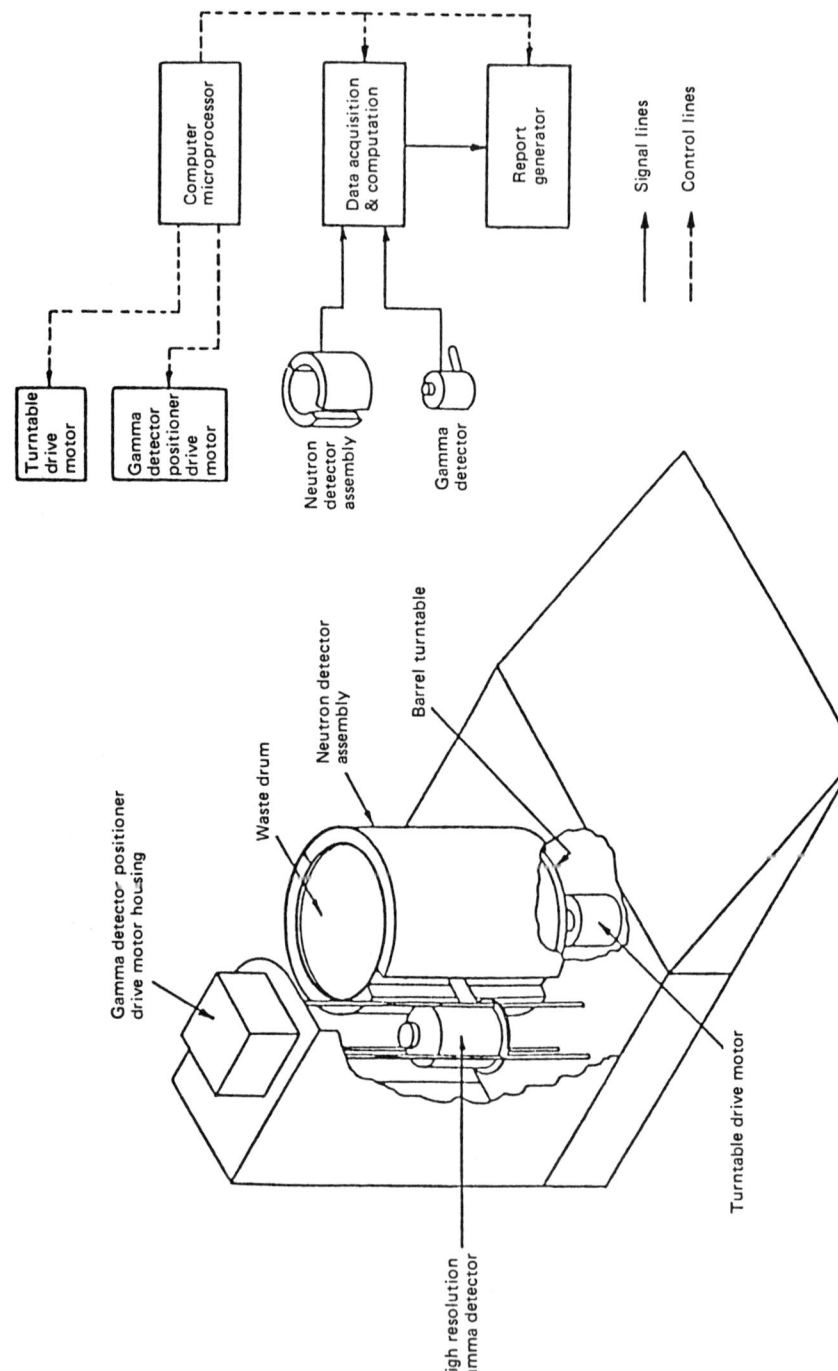

Turntable drive motor

Gamma detector positioner drive motor

Computer microprocessor

Data acquisition & computation

Report generator

Neutron detector assembly

Gamma detector

Signal lines

Control lines

Gamma detector positioner drive motor housing

Waste drum

Neutron detector assembly

Barrel turntable

Turntable drive motor

High resolution gamma detector

FIG. 3. A segmented gamma scanner fit with neutron counters and capable of simultaneously measuring gamma ray emitting radionuclides, ^{90}Sr, and transuranics.

28

At the Savannah River Plant in the USA a high throughput waste assay system with a computer controlled gamma spectroscopy unit is used to quantitatively measure many radionuclides in waste, including waste in heavily shielded shipping casks [44]. The computer program corrects for factors such as package geometry and shielding, undetectable radionuclides (through scaling factors) and contamination on reusable containers. This system is still under development. The area most in need of further work is that of scaling factors. Efforts are continuously under way to optimize both the factors used and the ability to efficiently incorporate them in the control software.

At the Chalk River Nuclear Laboratories in Canada a prototype waste assay system is being designed for use in a high throughput process environment for reactor and laboratory wastes [45]. The monitoring process will be activated automatically by insertion of waste items into the unit. After a preset total gamma ray assay of a few seconds the computer will decide if the activity on the package is high enough to start collection of a gamma spectrum for 60 seconds. In this way all waste will have a total gamma assay, but only the highest waste will be assayed by gamma spectrometry.

In France, at the La Hague plant, a system for measuring the activity content of 200 L drums containing ^{137}Cs immobilized in an epoxy resin matrix is fully operational. By means of a special gantry, the drum is rotated in front of a sensitive Geiger-Müller detector and the resulting signal is converted into total acticity in the drum. In a dismantling unit for TRU equipment at the same plant, four vertical ^{3}He counters placed around the drums of waste are used to measure, and with the help of a desk-top computer to calculate, the ^{239}Pu, ^{240}Pu and ^{241}Am content of the waste. Subsequently the drums are moved to an adjacent unit and filled with concrete to immobilize the waste.

In Finland [46] a system for measuring the activity content in solid low level waste from nuclear power plants has been developed. The total gamma activity of waste in containers is measured using a sodium iodide detector and a rotating and elevating mechanism for the container. The NaI monitor was selected as being best suited for monitoring total gamma activity because of its high efficiency. It is felt that the activity level of pure beta emitters, especially ^{90}Sr, can be conservatively estimated from the measurement of ^{134}Cs or ^{137}Cs activity.

In the United Kingdom an integrated suite of waste assay instruments for TRU has been in operation at Dounreay since 1978 [47–49]. This instrumentation is designed to handle the waste streams arising from normal operation of a fuel reprocessing plant. Development of instrumentation for use in the decommissioning of facilities which have handled large quantities of fissile material is being carried out [50, 51]. Such facilities need special care in decommissioning because of the risk of a criticality accident.

Elsewhere studies have been proposed to develop reasonably straightforward techniques for assessing the radionuclide inventory of packages containing waste from decommissioned facilities [52]. It is proposed that an external gamma ray spectroscopy system be used to identify the radionuclides present and determine the quantities of each radionuclide.

Commercial units to segregate active and inactive low level non-TRU waste streams so that the inactive materials can be released for unrestricted use or for disposal in landfill sites have been developed by several companies [53, 54]. One unit uses rugged liquid scintillation detectors for measuring gamma levels. This unit also has separate beta sensitive monitors and a belt conveyor system to carry the waste.

The application of such on-line systems for characterizing decommissioning wastes should prove to be cost effective and give greater assurance that active waste will not be inadvertently sent to an inactive landfill site.

5.5. Waste management

The general methods for treating, conditioning, handling, storing, transporting and disposing of radioactive wastes arising from decommissioning will be in general similar to those used in other parts of the nuclear industry. Since most methods have been well covered in other IAEA reports [55–63], they will only be briefly reviewed here. However, owing to the specific nature of certain decommissioning wastes, special consideration may be necessary in certain areas.

The main objective of any waste management strategy is to guarantee the safety of all the waste operations. Detailed consideration must be given to the types of wastes, packaging processes, containers and transport. The waste forms and packaging have to comply with national transport regulations and with the acceptance criteria at waste disposal sites.

In the following subsections the elements of a waste management programme for decommissioning are briefly reviewed, highlighting any differences between decommissioning wastes and those arising from other parts of the nuclear industry.

5.5.1. Sources and types of wastes

Some waste materials arising from the decommissioning of nuclear facilities will be radioactive as a result of activation and/or contamination. However, in general a large part of the waste will be inactive.

Most activated material is contained within the reactor vessel and its internal components as well as in the biological shielding which surrounds the vessel. Typically, these components contain materials such as steel, aluminium, reinforced concrete, graphite and zirconium alloys. A list of the radionuclides associated with the activated waste is calculated using the analytical techniques described in Section 5.2.

Contaminated materials arise from the decommissioning of all nuclear facilities such as nuclear power plants, fuel processing plants and fuel reprocessing plants.

The process equipment and components used to contain the process material (whether it is reactor coolant or reprocessing liquids) become contaminated with fission products, activation products and alpha emitters, along with other parts of the facility if there are any liquid, gaseous or powder leaks.

Contaminated liquids can result from the decommissioning of a facility, for example the liquid wastes arising from the decontamination or flushing of systems. The types of radioactive contaminants in the liquid are dependent on the type of facility being conditioned and the exact location of the waste stream.

Inactive solid and liquid wastes also arise from the decommissioning of nuclear facilities. If appropriate segregation processes are available, the volume of active waste requiring treatment could be reduced significantly. Typically, non-contaminated solid waste materials can include items such as piping, pumps, tanks, duct work, structural equipment, and electrical equipment. Inactive liquid and solid wastes can be disposed of using conventional methods.

5.5.2. Estimation of waste volume

A volume estimation of the waste material by type (activated, contaminated, alpha bearing versus non-alpha bearing) in relation to the methods and processes available for their conditioning is required.

An accurate estimate of the volume of waste materials should comprise the following activities:

(1) Classifying facility systems and structures with respect to activity (activated, contaminated, non-contaminated, alpha bearing versus non-alpha bearing, etc.). This characterizes the type of waste that will be generated as well as the treatment, handling, packaging and disposal requirements;

(2) Developing a detailed mass/volume inventory of facility systems and structures;

(3) Determining the quantities and volumes of compactible and incinerable solid wastes including items generated during decommissioning;

(4) Determining the quantities and volumes of solid wastes which cannot be compacted or incinerated. As this waste category has a large impact on the technical equipment required for handling and conditioning, an accurate determination is required;

(5) Determining the volume of liquid wastes. The volume of liquid wastes generated during decontamination and flushing operations will largely depend on the type of facility and its representative contaminants, the number of decontamination steps, and their efficiency;

(6) Determining gaseous and aerosol wastes. Aerosols containing finely divided radioactive materials result from cutting and abrasive surface cleaning methods. Some cutting and cleaning methods produce large volumes of toxic smoke and fumes. Contamination control coupled with filters in the ventilation streams should be adequate to collect and retain the particulate material.

Small quantities of tritiated water vapour may result from decommissioning operations. If necessary, removal of the tritiated water vapour from the ventilated air can easily be accomplished [55].

5.5.3. Requirements for waste treatment

The wastes should be treated according to the types and concentrations of radionuclides present and to the criteria of the waste disposal site.

The choice of conditioning processes will depend on a variety of parameters including:

— the physical, chemical and radiological properties of the waste as well as the ability of the radionuclides to migrate in the disposal environment;
— the type of treatment processes available, for example compaction, incineration, bituminization, etc. The location of the equipment to do the job, for example: are the conditioning facilities on-site or at a nearby location?
— the transportation, storage and disposal alternatives available;
— economic considerations, for example the cost of disposal compared with the cost of treating the waste.

Where possible, the waste is pretreated to provide appropriate preparation of the waste to facilitate subsequent waste treatment steps. Pretreatment steps could include:

— administrative steps including the documentation of the details of the waste for accountability and operational purposes;
— segregating and sorting waste to classify it for suitable treatment;
— packaging in containers (bags, drums) suitable for transport to the treatment area;
— decontamination;
— intermediate decay storage to allow decay of short lived isotopes to facilitate subsequent treatment steps.

5.5.3.1. Treatment of solid wastes

Solid low and intermediate level wastes are generally segregated into incinerable, compactible and non-compactible wastes [57].

Incineration of combustible wastes gives a large volume reduction, for example 100:1, and produces a stable waste product (ash) which can be readily immobilized using a variety of methods employing different matrices such as contrete and bitumen. Numerous types of incinerators are in use or under development for radioactive wastes, for example: excess or controlled air incinerators, fluidized bed incinerator, acid digestion, etc.

Compaction is the least expensive and easiest to operate volume reduction process, although it only produces reductions of about 6:1. High pressure compactors can give somewhat better reduction factors. Compaction units are also amenable to automation which can improve operational efficiency and radiation protection aspects.

Treatment of solid wastes that are not combustible or compactible generally requires some segmentation (see Annex B) so that standard types of disposal containers can be used. The extent to which segmentation is required depends on the capacity of the transport methods and the size or weight restrictions that exist at the disposal site. Where the cost of disposal is very high, the melting of metals could be considered to reduce the volume considerably. The cost effectiveness of this approach can be improved if the metal can be used as additional shielding for other waste. Where possible, it may be considerably cheaper to remove, ship and dispose of large units, for example the pressure vessel, in one piece [64].

Because of the radiological characteristics and the special requirements for the disposal of alpha bearing wastes, there may be an incentive or even a requirement to segregate waste streams during processing and packaging.

5.5.3.2. Treatment of liquid wastes

A large number of processes are available to reduce the volume and to immobilize the low and intermediate level liquid wastes which will arise from the decommissioning of nuclear facilities [58]. Processes to reduce the volume of the radioactive liquid wastes include: filtration, flocculation, precipitation, ion exchange, evaporation, etc. However, certain liquids used during decontamination, for example concentrated acids, may require special treatment processes.

The concentrated radioactive residues contained in the ion exchange media, filters, or concentrator bottom liquids require conditioning to immobilize those residues. Matrix materials that are used for immobilization include cement, bitumen, a variety of polymers, glasses, etc. [56].

5.5.3.3. Recycle and reuse of decommissioning wastes

During the decommissioning of nuclear facilities quantities of valuable metals and equipment may become available for recycle or reuse, provided that the activity on or in them can be reduced to acceptable levels. Work is currently in progress within the international organizations to define the basis for establishing suitable criteria for unrestricted release of materials and equipment and for applying these criteria to actual waste management and decommissioning cases (Section 5.3).

Some countries have already released items for unrestricted release on a case by case basis. For example, 900 Mg of metal scrap from the Würgassen nuclear power station is to be decontaminated and reused as normal scrap metal [65]. The NS Otto Hahn was decommissioned to Stage 3 by removing all the nuclear parts and cleaning up any residual activity. The ship can now be used as a normal ship [66]. Another method is to remelt the scrap. The reduction in activity of the final product results from the decontamination processes before melting and by partition of the radionuclides to the melt, slag and dust during the melting operation. The CEC is investigating remelting as a promising method for conditioning steel waste with the purpose of volume reduction, immobilization of radioactivity and possible recycling of the steel [12, 67].

During the decommissioning of a nuclear facility the wastes are usually segregated as much as possible according to the final method of disposal or reuse. There are generally five categories of material:

(a) Higher level waste, which usually goes to some form of high integrity repository
(b) Low and intermediate level waste to appropriate repository such as shallow land burial
(c) 'Inactive' waste to landfill
(d) 'Inactive' metals for remelt and recycle
(e) 'Inactive' equipment for reuse.

Suitable criteria and instrumentation must be available to segregate (a) and (b) from (c), (d) and (e). This is particularly important for item (c), which usually represents a very large portion of the waste. In setting the critera for the management of waste categories (c), (d) and (e), it may be desirable to have one criterion for all three wastes rather than three criteria if this is possible.

Another important factor which must be considered in the unrestricted reuse or recycle of material is the economic impact [68]. It is beyond the scope of this report to do a detailed economic study of this approach to waste management. However, the factors which should be considered in assessing the economics of recycle will be briefly discussed.

The cost savings in recycling or reusing materials are relatively clear:

— there are no disposal costs for the item
— the scrap value.

Both factors depend on the circumstances in a particular country. For example, the cost of the least expensive disposal method can range from one to several thousand dollars per cubic metre. Even at the lower value the cost savings can be significant. The scrap value depends on the individual item.

On the other hand, the costs to reclaim the scrap can be significant and include:

— material and labour costs to decontaminate the materials;
— costs of treating and disposing of the wastes arising from decontamination;

— cost (labour and man-sievert) of doing the extra monitoring to select the items for recycle and ensuring that the pieces are below the selected limits. The cost of this item will increase as the level of the acceptable limits for release decreases since the main problem with very low level waste are the measurability of the inherent activity and verification of the measuring techniques used;
— cost of treating and disposing of the radioactive slag and dust from the remelt furnace;
— possible eventual decontamination of parts of the remelt furnace if it becomes contaminated.

It is obvious that there are many factors to be evaluated when considering the recycle/reuse of equipment and material. It is recommended that, in addition to the criteria for unrestricted reuse/recycle, more work be done on the economics and safety.

5.5.4. Transport of conditioned wastes

Waste should be packaged in a safe manner according to national regulations for transportation and criteria of the disposal site. Transport of conditioned wastes to the disposal site is performed using conventional means via highways, railways or waterways. The mode of transport utilized is influenced by the type of transport access and handling equipment available at the packaging site and at the selected disposal site, and by the physical size of the package unit.

For large scale decommissioning projects, e.g. a nuclear power reactor, the logistic problems of moving many thousands of cubic metres of waste are likely to be significant and the total impact on existing systems should be carefully assessed.

The sizes and weights permitted in shipments are governed by regulations issued by the appropriate national and local regulatory agencies. Protection of the public from harmful radiation is governed by the appropriate national regulations, which are based upon international recommendations [59].

5.5.5. Disposal of wastes

The method for disposal of radioactive wastes is governed by applicable national (and international) regulations, by the availability of appropriate disposal facilities, and by the need to achieve an optimum cost–benefit ratio for accomplishing the disposal. The type and specific activity of the radioactive material present in the waste are some of the most important factors in selecting the disposal method. Other important factors are the size of the package and the difficulty in handling the package during disposal.

The principal methods for disposal [60–63] are shallow ground disposal, emplacement in rock cavities or repositories in deep geological formations, as well as possible deep sea dumping [69]. Shallow ground disposal is generally employed for low and intermediate level radioactive wastes. Sea dumping may also be used for certain types of low level relatively short lived radioactive wastes. Rock cavities can potentially be used for all kinds of solid low and intermediate level waste. Disposal in deep geological formations is envisaged for high level wastes with significant quantities of long lived radionuclides.

The choice of the disposal method is dependent on conditions in the country and on many other factors specific to the disposal system to be developed. Generally, shallow ground disposal and rock cavity concepts appear to be the most viable for disposal of decommissioning waste.

5.6. Decontamination

In this section the topic of decontamination is only briefly reviewed since it has been covered extensively in other documents [6, 9, 14, 70].

During decommissioning, decontamination is used to reduce radiation fields by removing some of the fission and activation products contained in deposits, oxidation films and dust in the facility to minimize radiation exposure to the decommissioning staff and the public. Other reasons for carrying out decontamination include recovery or reuse of the material and compliance with the repository requirements. The methods used are generally based on cleaning techniques which have been well established for decades in non-nuclear plants. However, these techniques usually have to be adapted to the deposits in a particular plant to obtain optimum decontamination efficiency and to take into account the very large areas which have to be decontaminated.

Very early in the selection process it is important that a cost–benefit analysis be done to see if it is actually worth decontaminating the facility or to see if a mild decontamination at low cost is more advantageous than a severe decontamination at a higher cost.

If the decision is made to decontaminate a specific nuclear facility before dismantling it, a great many factors must be considered to achieve good decontamination efficiency. This analysis is usually accompanied by extensive experimental work on samples selected from the facility before the final choice is made.

In the following sections the factors which must be considered in doing the analysis are briefly reviewed along with the types of decontamination processes which are currently being used in nuclear installations.

5.6.1. Selecting a decontamination process

To achieve a good decontamination factor (DF), a decontamination process must be designed for site specific application taking into account a wide variety of parameters, some of which are listed below:

- The type of plant and plant process: reactor type, reprocessing plant, etc.
- The operating history of the plant
- Type of material: steel, Zircaloy, concrete, etc.
- Type of surface: rough, porous, coated, etc.
- Type of contaminant: oxide, crud, sludge, loose, etc.
- Composition of the contaminant: activation products, fission products, actinides, etc.
- External or internal surface to be cleaned
- The decontamination factor required
- Destination of the components being decontaminated: disposal, reuse, etc.
- Time required for application
- The proven efficiency of the process for the contamination in the facility
- The type of component: pipe, tank, etc.

Other factors which are important in selecting the method but which do not affect the DF are:

- Availability, cost and complexity of the decontamination equipment
- The need to condition the secondary waste generated
- Occupational and public doses resulting from decontamination
- Other safety, environmental and social issues
- Availability of trained staff
- Extent to which the plant needs to be decontaminated to achieve acceptable conditions for decommissioning
- Salvage value of materials which would otherwise be disposed of
- Extent to which the facility must be modified to do the decontamination: isolate systems, enclosed and ventilated spaces, etc.

In summary, the decision whether to proceed with decontamination and the final process selected will depend on the best overall balance of the above factors to minimize the overall impact of decommissioning and the net detriment to man.

5.6.2. Decontamination processes

A large number of decontamination techniques and a large variety of chemical mixtures have been developed over the years to assist in removing contamination from metal and concrete surfaces. Currently research programmes are being carried out in many countries to improve or to develop specialized decontamination processes for site specific application.

In general, decontamination processes are classified as chemical, mechanical, and other.

5.6.2.1. Chemical decontamination

Chemical decontamination methods are generally required to clean contaminated metal surfaces. The processes are generally classified as high or low concentration methods (with the dividing line at about 1% concentration of reagent), electrochemical decontamination and other chemical processes such as foams, gels and strippable coatings.

In general, the high concentration processes are used for limited volume subsystems or for components which can be immersed in a tank. Although the time of application is usually relatively short, these processes generally require repeated circulation and water flushes. The high concentration processes have higher decontamination factors but are more corrosive to the base metal and generate large volumes of wastes. The higher corrosion rates might not be a problem if the facility is to be decommissioned to Stage 3.

The low concentration processes can be applied to entire reactor coolant circuits, or to components for tank immersion. In these processes the inventories of chemicals are low and they can be regenerated during the decontamination process with ion exchange systems. The process results in low corrosion of the base metal and low waste generation (expended ion exchange resins). However, longer circulation times are required and DFs tend to be lower than for high concentration processes.

Electrochemical or electropolishing decontamination methods are variants of chemical processes and consist of dissolving the surface of the object by immersing it in an electrolyte so that it becomes the anode in an electrical circuit. The electropolishing process can be readily adapted for remote operation, has high DFs, short process times and low waste volumes. However, these techniques are fairly expensive, labour intensive, difficult to apply to complex geometries, and generate large volumes of hydrogen gas.

5.6.2.2. Mechanical decontamination

Mechanical decontamination methods involve the physical removal of the contaminants. These methods may give rise to large volumes of dust, aerosols, fumes and solid or liquid wastes which must be contained and conditioned to avoid spread of contamination and exposure of personnel, particularly by inhalation of dust or aerosols. Some processes under this heading are:

- **Vacuum cleaning**: applicable to non-adherent particulate contamination on accessible surfaces. The process is cheap, equipment readily available and the waste is easily handled. However, personnel exposures could be high and DFs are usually low;
- **Washing, swabbing or scrubbing with or without solvents**: an inexpensive domestic process that is applicable to components of all sizes and is useful for loose or soluble contaminants. However, spread of contamination is easy and large liquid or solid waste volumes can be produced. Personnel exposures may be high;
- **Water–steam jetting**: a process using high pressure liquids to clean a wide variety of surfaces. The equipment is readily available and can be adapted to remote operation. Good contamination factors are obtainable;
- **Abrasive jetting**: an efficient decontamination process for metals and concrete. A wide variety of abrasives such as sand or carborundum are carried in the high velocity air, water or steam to clean the surface;
- **Spalling**: this process uses mechanical forces transmitted through a bit with expanding wedges placed in a predrilled hole to create a crater to remove layers from concrete or masonry surfaces;
- **Flame spalling**: uses heat induced differential expansion to spall material from concrete surfaces. Application as for spalling. Penetration is less than 5 mm per pass;
- **Scarifying**: similar to spalling but offering deeper penetration per pass.

5.6.2.3. Other decontamination processes

Within this category can be included:

- **Ultrasonic cleaning**: The method is normally confined to the cleaning of tools and other small items, particularly precision parts. The advantages are short process time, no secondary wastes and high DFs for certain applications. Decontamination factors are relatively low for complex components.
- **Freon process**: This process is still under development and is particularly suitable for electrical equipment and cables. The freon, an organic solvent, acts by mechanical dislodgement and/or chemical dissolution and quickly evaporates leaving a dry, residue free surface.
- **Melting**: This method is totally destructive and is effective only for contaminants that are more soluble in the slag than in the molten metal.

Decontamination processes are undergoing continuous development and experimentation with different solvents, combinations, and sequences of application. Lists of standard solutions and proprietary mixtures, decontamination efficiency and fields of applicability are available and should be referred to for up to date state of the art information when undertaking a specific decommissioning project.

5.6.3. Decontamination of sites

The last process in the decommissioning procedures is usually the cleanup of the site for unrestricted use. In case of facilities in which the radioactivity in the process system has been essentially kept within the buildings the cleanup of the site is not a severe problem. Examples of such situations are reactors, fuel fabrication facilities and reprocessing plants.

However, for certain facilities, for example mining and milling sites and phosphate production plants, the incoming materials and radioactive residues are stored outside the plant buildings. In these cases the cleanup of the site is much more difficult.

On sites with contaminated areas the first step in the cleanup process is to carry out a detailed contamination survey delineating the area and depth of contamination, the radionuclides present and the physical form of these nuclides (particles, homogeneous mix, etc.). At this stage a detailed knowledge of the facility's operating history, with particular reference to radiological incidents, should be made available to the decommissioning project. The second step, if unrestricted use of the site is required, is to physically remove the radioactive residues and any contaminated earth in which the radioactivity exceeds the approved limits.

5.7. Disassembly techniques

The dismantling of nuclear reactors and other facilities contaminated with radioactivity generally involves the segmentation of metal items such as reactor vessels, tanks, piping and other components. In addition, in most facilities to be decommissioned demolition of concrete components and/or scarification of the surface to remove active or contaminated areas are required.

The cutting processes applicable to metal components include: the arc saw, spark erosion, plasma arc cutting, oxygen burning, thermite reaction lance, explosive cutting, hacksaws, guillotine saws, band saws, abrasive cutters, circular cutters, and pipe crimpers.

Concrete removal processes include: controlled blasting, wrecking ball, rams, flame cutters, rock splitter, sawing, drilling, explosive cutting, drill and spall, scarifying and sand/water/shot blasting.

Since some of the components which are to be cut or scarified are highly radioactive, the equipment used for the above techniques must be operated remotely in many cases. A wide range of technology and specialized equipment has been developed in connection with the maintenance, repair, refurbishment and replacement tasks in nuclear facilities. This development will proceed independently of decommissioning requirements.

Advances in automated and computer controlled robot arms have been achieved and much of the technology can be applied to improving the remotely controlled equipment used in decommissioning and other nuclear activities [71]. The application of robotics to decommissioning activities is already in progress.

More details on the various segmenting and scarifying techniques mentioned above and on remotely controlled equipment and robotics applied to decommissioning are given in Annex B (Disassembly techniques) and Annex C (Remotely controlled equipment for decommissioning).

5.8. Long term integrity of buildings, structures and materials

One of the major factors to be considered in the choice between prompt versus delayed dismantling of a nuclear facility is the stability of the buildings and barriers containing the radioactive materials. People and the environment must be protected from inadvertent release of radioactivity owing to deterioration of barriers during long term storage.

Studies are being done under contract for the Commission of the European Communities [67, 72] on the behaviour of reactor buildings and structures over a prolonged period of time. Preliminary results showed that the in situ storage of shut down facilities is a feasible option. However, technical surveillance and maintenance may be required.

For water cooled reactors there apparently will not be any inherent difficulties in leaving the main circuits drained and filled with dry air for at least a few decades.

For gas cooled reactors the study suggests that it would be desirable to seal off the reactor pressure vessel from the main gas circuits forming smaller sections to limit spread of contamination through natural convection.

It was emphasized that the major parts of auxiliary plant items such as electrical circuits, ventilation and lifting devices have a life span in the range of 25 to 35 years. These items may have to be requalified or replaced to carry out dismantling if decommissioning is delayed past about 30 years.

To back up these preliminary studies it was recommended that experience gained with plants which have already been shut down should be continuously analysed.

The Central Electricity Generating Board in the United Kingdom is examining the long term integrity of reinforced structural concrete, especially biological shields. It is possible that deterioration of such structures may occur if left for a 100 years after 30 to 40 years of operation, especially if the concrete has been subjected to heat and radiation.

5.9. Decommissioning cost estimating and financing

Cost estimates for decommissioning are essential to plan an economically sound decommissioning programme and select a practical mechanism. Published estimates of decommissioning costs for comparable plants prepared by different organizations vary widely. In some cases the differences can be associated with different policies, contingency allowances, work scopes, labour rates and money values due to inflationary considerations. Lack of a common methodology may be a further source of disparity. However, the greatest source of incertitude lies in the fact that no large decommissioning project has yet been accomplished.

One approach to standardizing decommissioning cost estimating is to provide a detailed activity by activity cost estimate using the 'building block' concept; that is, the whole decommissioning programme including construction of needed facilities and plant, waste treatment and disposal is broken into a series of discrete and measurable work activities. The breakdown should be carried to a level of details such that measurable events are repetitive. The summation of all costs in a comprehensive and consistent manner, supported by defensible reference sources for cost data, provide a reasonable basis to estimate current and future costs. These cost estimates can then be updated periodically to reflect changes due to inflation, regulations or technology so as to provide adequate funds for decommissioning.

Of course the success of using the building block approach to estimate decommissioning costs depends to a great extent on how much of the work can be expressed in this manner, how accurate the estimates are to do the unit job, similarity of facilities, uncertainty in waste management costs, etc. It is quite possible that reasonably good estimates will be achievable using this approach when much more decommissioning experience is built up, especially with similar units in the same country and with well defined waste management costs. On the other hand, if the unit cost factors are not very well defined for a particular facility in a particular country, large contingency factors would be needed to cover the uncertainties in costs. Obviously, if a job has not been done before, then unit cost factors could not be used and cost estimates would have to be done in the regular way.

If there are long delays between phases, the forecasting of funding requirements will become even more uncertain owing to a large number of factors such as potential regulatory changes, social and industrial influences, and personnel turnover and retraining. Furthermore, the costs of maintenance, surveillance and security need to be provided for.

Since the costs of each decommissioning alternative have a strong influence on the selection process, cost estimates must be made on a site specific basis each in its own context to reflect the site conditions and desired disposition of the facility, but attention must be given to management of costs and technical uncertainties.

Funds required to implement decommissioning are also required for the immediate post-operating phase and the dismantling phase. The former phase may include the costs of planning and engineering. The second phase will include preparatory work, then dismantling, demolition, transportation and disposal. In the meantime, costs for maintenance, surveillance and security of the installation will have to be met.

It must be recognized that decommissioning is not the only contingency with respect to which financial provisions shall be made. Situations in which a large power plant is forced to a premature final shutdown as a consequence of a serious failure must be considered also in their financial aspects.

The methods of financing decommissioning, as well as of covering these risks, are largely dependent on the size of the national nuclear programme and the size of the operating organization with respect to the size of the nuclear facility.

This report deals only with the costs associated with a planned decommissioning at the end of the planned operational life of the facility, and with the relative financing.

The availability of funds to pay for all the costs of decommissioning must be ensured by establishing a suitable financing plan. Several methods for funding have been identified and include prepayment at commissioning (facility startup), externally or internally held sinking funds, insurance fund and payment at decommissioning.

5.9.1. Elements of decommissioning costs

A major part of decommissioning costs are more similar to the costs of a construction project than the operating costs of a facility, because of the need to design, engineer, procure and erect specialized and dedicated structures, plants and equipment.

The basic elements of all cost estimates include labour, materials, equipment, energy and services. One approach to estimate costs is to classify the types of costs into three categories; (1) activity dependent costs; (2) time dependent costs; (3) special costs. Each category includes some or all of the basic elements identified earlier, but this simplifies the estimating process.

Activity dependent costs are those directly related to engineering, development, preparatory works, decontamination, demolition, packaging, shipping and disposal. They include all labour, materials, equipment and services (shipping and disposal) associated with the 'hands-on' activities.

Time dependent costs are those associated with project management, administration, routine maintenance, radiological, environmental and industrial safety and security. They are not directly assignable to any one activity, but continue for the duration of the decommissioning programme.

Special items costs are generally one time costs such as licences, permits, and heavy equipment purchases.

5.9.2. Cost estimating guidelines

The preparation of cost estimates using the above approach relies on the development of 'unit cost factors'. Each repetitive event such as cutting pipe, segmenting vessels, demolishing concrete, transporting wastes and disposing of wastes including labour, equipment, materials and services is individually cost estimated. The unit cost factors are then expressed in terms of the cost per cut, cost per cubic metre demolished, cost per trip or cost per metre of burial, etc.

However, it is important to note that these costs represent only part of the overall costs, a major part being the cost of construction and preparation work. If several plants of the same design have been decommissioned, then it may be possible to develop cost factors for some of the construction and preparation work, otherwise large contingency factors would have to be included.

Sources of information for developing costs include recorded experience, estimating handbooks and equipment catalogue performance data. Calculations of activity dependent costs include set-up time, operating time, manpower required, consumables, support services and energy consumption.

Preparation of a facility decommissioning cost estimate consists of the following steps:

— Work sequence development
— Activity dependent costs
— Programme schedule development
— Time dependent costs
— Total programme costs.

These steps are discussed in more detail below.

5.9.2.1. Work sequence development

The detailed work sequence will include all preparatory and implementation steps of the decommissioning programme including planning, licensing, detailed engineering, work performance and release of site or facility. This detailed description of the whole decommissioning programme would be laid out in a logical time related sequence of series and parallel work activities.

The preparation steps include all planning and engineering tasks such as performance of radiation surveys, calculation of activation and contamination inventories, performance of engineering studies, preparation of a decommissioning plan, preparation of major activity specifications and descriptions, design of special tools, and preparation of detailed decommissioning procedures.

The implementation steps would include all physical tasks such as decontamination, equipment removal, structure removal, radioactive waste packaging, and shipping and disposal.

The final step would include demolition of all remaining non-radioactive structures, if desired, and final site and facility restoration or preservation.

5.9.2.2. Activity dependent costs

Each work activity is then estimated as to cost and the time required for completion. The cost estimate is derived from calculations or from actual field experience. Costs derived through calculation based on a performance parameter (speed of cut, for example) must be modified by allowance for operator efficiency and work area conditions. Work output in radiation and/or contamination areas will be reduced dramatically from theoretical output owing to the difficulty of working in protective clothing or respirators and other radiation control measures and therefore costs must be estimated on a case by case basis according to the conditions likely to be experienced.

The basic elements of cost for any task or subtask in a decommissioning work sequence are labour, materials consumed, equipment, energy and service. Sources of information on these costs are included in Refs [26, 73—77].

The activity dependent costs should include costs relative to the construction of buildings and structures necessary for access to, dismantlement and removal of components, special tools for such operations, facilities for performing decontamination and treating secondary wastes, etc. These costs could be higher than the direct costs of decontamination, demolition and removal of plant and structures.

5.9.2.3. Programme schedule development

At this point a detailed programme schedule can be developed based on the calculated work activity durations and the sequential relationship between activities. The possibility of performing in parallel different activities should be considered to save time and to optimize the utilization of resources.

5.9.2.4. Time dependent costs

Time dependent costs are then calculated as a function of work phase duration. These costs include such items as project administration, insurance, site security, health physics support, quality assurance and certain equipment rentals when their use is common to many activities and are obtained by estimates made over the expected duration of the individual activities.

5.9.2.5. Development of total programme costs

The sum of the activity dependent, time dependent and special items costs represents a 'best estimate' of the actual costs of decommissioning. However, there will always be a range of variability as influenced by programme assumptions and cost assessment accuracy.

All engineering activities involve uncertainties and it is prudent to allow some contingency for unforeseen circumstances. This applies to decommissioning as well, but decommissioning may involve greater uncertainties than construction work for the following reasons:

(a) The extent of the work may be uncertain in that activation and contamination may not be defined fully at the start of the work. The nature of the contamination may also not be defined until the work has begun;

(b) Working conditions, which affect work rate, may turn out to be worse or better than predicted and provision must be made in the work programme to accommodate this;

(c) Further, regulatory changes may have important consequences for working practices including waste management, as may social and industrial factors.

Studies by the AIF [78], the NRC [73] and the CEGB [79] include an accuracy analysis of the cost estimates.

In the international arena it is difficult to express decommissioning costs in a readily comparative manner. The differences in site specific factors coupled with differences in scope of selected decommissioning stages make comparisons from country to country widely different on an absolute basis. The various monetary units further complicate the issue of trying to establish cost estimate benchmarks (comparative bases). Therefore, available studies should be used only as a reference, not necessarily applicable, as a whole or in part, to any new project.

5.9.3. Financing methods and approaches

Costs for decommissioning nuclear facilities should be estimated well in advance of plant shutdown to determine how much funding will be needed and how to pay for it. The cost units may be estimated by the methods described in Section 5.9.2 and should be reviewed periodically to adjust for economic, political and technological changes. The funding methods for each type of facility depend on the preference of the owner organization within the framework of national laws or governmental decisions. The actual funding practices vary widely both nationally and internationally.

The two primary objectives in funding for decommissioning are: a high degree of assurance of the availability of funds and low cost of providing that assurance. The degree of assurance is a measure of how effective the funding method is in providing funds when needed. Since decommissioning is a phase of the life of the facility, it is generally agreed that all decommissioning costs, including financial charges, are to be considered as a part of the overall revenue costs of the facility, unless national policies establish a different position.

The funding methods and degree of assurance of each method are described below.

However, it is not intended to propose to Member States any particular course of action in this field. It is up to each project to decide which approach is appropriate to its needs.

5.9.3.1. Prepayment

In this method cash or other liquid assets that will retain their value for the projected operating life and subsequent decommissioning of the nuclear facility are deposited before startup into an account segregated from owner or licensee assets and outside its administrative control. Prepayment can be in the form of a trust, certificate of deposit, government security, escrow account, or government fund.

Prepayment provides a good assurance that funds will actually be available since the necessary funds are deposited at startup. Periodic review and adjustment of the fund will likely be necessary over the period of licensing because of uncertainties in cost estimates and changes in inflation and interest and other economic and political factors.

5.9.3.2. External sinking fund

The external sinking fund requires that a prescribed amount of funds be set aside in an account at fixed regular intervals over the life of the facility, such that the funds plus accumulated interest would be sufficient to pay for decommissioning costs at the time termination of operation is anticipated. The account would be segregated from owner or licensee assets and outside licensee control. Types of accounts could be similar to those described above for prepayment.

The external sinking fund is held outside the licensee's assets and control and would not be vulnerable under most likely trust arrangements if the licensee went bankrupt. On the other hand, in the event of premature decommissioning, there would be a greater likelihood than with the prepayment method that insufficient funds had been accumulated. This situation would be mitigated if the fund was either structured so that higher payments were made earlier in a facility's life, or coupled with a deposit, insurance or surety.

5.9.3.3. Internal reserve

This approach usually uses negative net salvage value depreciation which allows estimated decommissioning costs to be accumulated over the life of the nuclear facility. In this method the funds are not segregated from the company's assets, rather they are invested in its assets. At the end of the life of the nuclear facility bonds are issued against these assets and the funds raised are used to pay for decommissioning. This approach can also take the form of a segregated internal reserve, which is similar to an external sinking fund, except that funds are held by the company.

Under normal circumstances the internal reserve would be similar to the external sinking fund in the pattern of funds set aside and should provide adequate funds if a nuclear facility is decommissioned at the end of its expected life. However, because it depends on financing internal to the licensee, the internal reserve is vulnerable to events or situations that undermine the financial solvency of a licensee. A bankrupt or financially troubled licensee would have difficulty in raising capital against its decommissioning reserve and even a segregated internal reserve fund may not be available to pay for decommissioning costs.

5.9.3.4. Insurance, surety bonds, letters and lines of credit and other guarantee
            methods

Insurance could be used to provide coverage for premature decommissioning expenses. An insurance type mechanism might also be used for all decommissioning expenses, including those planned under normal circumstances. The surety bond, credit methods, and other guarantee methods ensure that the decommissioning costs will be paid should the licensee default. The licensee would still be responsible for paying for decommissioning. The costs of guarantee methods, such as sureties, letters of credit, or insurance, would be in addition to normal decommissioning expenses.

5.9.3.5. Payment at decommissioning

This method of funding requires a capital expenditure at the time of decommissioning. It is typically the approach used by government agencies for government owned facilities. The cost for decommissioning is identified to the agency and the agency in turn requests a capital expenditure either for the entire programme, or for an annual disbursement

6. DECOMMISSIONING SAFETY

After the final shutdown of a nuclear facility and the subsequent removal of fuel and/or process material the potential hazard to the general public and the workers is reduced considerably below that present during the plant's operating phase. However, in the case of plants which handled fissile material the possibility that some residual material is held up in the plant must be given serious consideration during decommissioning and adequate precautions taken.

The safety considerations which must be taken into account during decommissioning are:

(a) Radiological safety when the original containment barriers are breached and radioactive waste is processed

(b) Industrial safety associated with the dismantling operations

(c) The radiation doses to the public as a result of routine or accidental releases of radioactivity

(d) The hazard to the workers and the public during the shipment of radioactive material to the disposal site.

Furthermore, to achieve a high safety standard during the decommissioning phase, a comprehensive quality assurance programme approved by the appropriate competent authority must be prepared and used.

An IAEA report, Safety in Decommissioning of Research Reactors [80], outlines in general terms the technical and administrative aspects relevant to nuclear safety in the decommissioning of research reactors. Conventional safety aspects during decommissioning are not covered.

7. FACILITATION OF DECOMMISSIONING

The decommissioning of small research, demonstration and power reactors and other fuel cycle facilities in the past (Annex A) has provided the nuclear industry with experience which has led to the identification of a number of ways to facilitate future decommissioning projects. The techniques summarized below to facilitate decommissioning could be included, built or implemented during one or more of the four phases of facility life: design, construction, operation, and decommissioning. When giving consideration to the introduction of any of the suggested methods into a particular facility, it is very important that a detailed cost–benefit analysis be done. In addition, optimization between plant construction, operation, and decommissioning costs should be emphasized at all times when considering factors to facilitate decommissioning.

The two primary objectives for including these techniques into a facility are to reduce both occupational exposures and the volume of waste generated during decommissioning. Some techniques will also benefit plant operation and maintenance.

References [81–84], on the costs and benefits of using the facilitation techniques discussed below, provide more details on some techniques which have been found to be useful in previous decommissioning work.

7.1. Techniques to reduce occupational exposure

Reductions in occupational exposures are possible for example by: planning work activities; providing adequate access to equipment to reduce the time that the workers remain in the high radiation areas; having well trained crews; using remotely operated tooling; placing fixed or portable shielding between the radiation source and the worker; decontaminating the process equipment if necessary before work starts.

7.1.1. Planning

One of the most direct and cost effective methods of reducing worker exposure during decommissioning or maintenance is through careful planning to minimize the time that workers are in the radiation zone. This planning should start at the design phase and continue through the startup and operational phases with the accumulation and updating of plant construction drawings, photographs and material specifications. These data provide an accurate description of the facility so that the most effective way of removing equipment and structures can be determined. During the last years of plant operation an accurate estimate of the inventory and location of radioactivity is required to assist in the planning of decommissioning to minimize worker exposure.

A wide variety of planning techniques can be applied during the design and construction phase to reduce occupational exposure during decommissioning and some are described below. Whether or not these are applied will depend on many factors such as cost–benefit and engineering practicality. The selection of materials with low concentrations of elements which form undesirable activation products, such as ^{60}Co, could be a major step in reducing worker exposure. Similarly, the use of layered or segmented concrete to ease demolition tasks is also worth considering in certain types of plants both from a dose reduction viewpoint as well as making segregation of active and inactive concretes easier. Suitable built-in ladders and walkways should be included, especially in areas of high radiation, to minimize the amount of temporary scaffolding that has to be erected during decommissioning. Adequate space should also be provided around equipment for easy movement of workers and/or tooling. Oversize doors or hatches should be installed to facilitate rapid removal of components or easy installation of shielding.

Planning during operation could include special emphasis on water chemistry control to minimize corrosion of components and the subsequent spread of corrosion contamination products, and maintaining good records of physical changes made to the plant.

Planning for the actual decommissioning includes the establishment of a well defined decommissioning programme including items such as: the purpose and status of the project; an assessment of alternatives; the organizations involved and their responsibilities; overall cost, schedule and technical approach; the management, engineering and specialized decommissioning techniques to be used; analyses of radiological and industrial safety aspects; and an assessment of socio-political aspects. Planning should also include the development of well trained work crews who are knowledgeable about radiological protection and the job to be done to keep exposures as low as reasonably achievable, social and economic considerations taken into account. The crews should be very familiar with each task to be performed, the equipment to be used and their assigned responsibility. In addition, workers who are not immediately involved in the task in progress should be trained to move out of the area of radiation until needed. The timing as to when the dismantling operation will start and the preselection of remotely operated equipment (see Section 7.1.2) can also have a large effect on the doses received by the operators. Usually, delaying decommissioning reduces occupational doses. Decontamination of certain facilities can also be a good method of reducing radiation fields and occupational doses. However, the value of decontamination must be balanced against the financial and man–sievert cost of doing the decontamination and of treating the waste arising from the process.

With respect to overall project planning, the sequence in which dismantling activities are carried out can have a large effect on the dose recieved by the crew. If the major activity in a component is due to relatively long lived radionuclides such as ^{137}Cs (half-life 30 years), then removal of the component early on in the decommissioning sequence can sometimes reduce exposures. On the other hand, if the high activity comes from short lived radionuclides, then delayed removal for one or two years could be an asset. If several similar facilities are to be decommissioned, sequential decommissioning may reduce doses because the crews and equipment are available and experienced.

7.1.2. Remote tooling and techniques

Remote system technology and tooling have been used in the past to reduce occupational exposures and to assist in decommissioning to segment items such as reactor vessels and internals, piping and concrete, to assist in the decontamination of equipment, to aid in the dismantling process, etc. For larger reactors and other nuclear fuel cycle facilities which will be decommissioned in the future exposure rates and contamination levels could be considerably higher and remote tooling will probably be more important.

Much of the technology used in earlier programmes is still applicable today. However, recent improvements and development of new equipment using computer controls and various types of sensors permit greater flexibility and creativity in their application. Although these developments combined with established processes promise to be a great asset to the decommissioning operator, much work needs to be done so that this technology can be successfully applied to decommissioning. A more detailed description of some of these techniques is given in Annex C.

7.1.3. Shielding techniques

Maximum use should be made of simplified shielding such as leaded blankets, lead sheet and, in some cases of extended exposure durations, lead brick to reduce occupational exposures during decommissioning. In certain cases rapid temporary shielding can be installed by stacking 200-litre drums and filling them remotely with water. The planner needs to weigh the benefits of installing extensive shielding against the exposure incurred during installation.

The use of remotely controlled vehicles to detect spots having high radiation fields and then to install shielding before man access is permitted should be seriously considered to reduce occupational exposures.

7.2. Techniques to reduce the volume of radioactive wastes

A variety of techniques are possible to reduce the volume of radioactive waste which has to be disposed of during the decommissioning of nuclear facilities. These techniques will facilitate decommissioning and include changes in design, operational procedures during decommissioning to reduce waste generation rates and volume reduction of the waste generated.

7.2.1. Design, construction and operational features

Activated or contaminated concrete are major sources of waste during the decommissioning of many types of facilities. This concrete includes the neutron activated biological shield surrounding the reactor vessel and floors and walls contaminated as a result of spills during operation.

Two design approaches have been considered for use in the construction of biological shields to minimize neutron activated concrete in the waste. The first consists of fabricating the entire biological shield from precast steel reinforced interlocking blocks held together with steel bands and bolts. In this design only the activated blocks need be removed for controlled disposal. The block approach eliminates the need for blasting or other dust generating methods which may cross-contaminate non-activated concrete, thereby increasing the waste volume. A second technique consists of fabricating the inner region of the biological shield from a plaster like material (possibly applied in layers) which could be easily demolished and would permit simplified removal of only the radioactive portions. The practicality and cost effectiveness of these approaches still needs to be proven, especially for power reactors.

Surface contamination on floors and walls can be minimized by using steel plates or gratings instead of concrete slabs. The steel flooring may be decontaminated to unrestricted release levels more easily than concrete, thus reducing the volume of waste. Furthermore, scarification of concrete floors may cause cross-contamination of clean floor or wall areas, thereby increasing waste volumes. However, if the concrete floors and walls in active areas are prepared with a smooth surface finish and protected with an epoxy or similar coating, decontamination of the concrete will be much easier and active waste volumes will be reduced from this source.

The design of rooms or cubicles for components containing contaminated fluids should include drip trays and floor curbs of sufficient capacity to contain the maximum credible spill or component rupture. The curbs should direct spills to floor drains with sufficient tankage to collect all waste. Care should be taken not to permit oil spills to mix with water based drainage.

7.2.2. Reduction of waste generation

During decommissioning operations major reductions in waste generation can be achieved through training and administrative controls, contamination control tenting and confinement, and decontamination of selected material and components to releasable levels.

Each worker should be trained in the need to minimize the waste generated in the tasks assigned to him. Generally, this consists of alerting workers to install contamination control tenting, contain spills, etc., and instructing workers not to mix non-contaminated wastes with contaminated wastes.

Contamination control tenting should be used wherever the potential for airborne contamination exists in cutting or grinding operations. Tents should be fitted with high efficiency particulate air (HEPA) filters with adequate air flow away from the worker to prevent ingestion. Plastic sheeting covered with absorbent pads should be used under all pipe cuts where there exists a potential for liquid spills to minimize the spread of activity from such a source.

Selected piping, components and tools may be decontaminated to unrestricted releasable levels by one or more techniques. In particular, components or tools having simple geometry and smooth surfaces are well suited for decontamination by wiping, washing, dry cleaning or by more sophisticated techniques such as electropolishing or vibratory finishing. Tools cleaned in this manner may be reused repeatedly. In each case the cost–benefit should be evaluated.

In addition to reducing the amount of contaminated waste from decommissioning, reduction in the volume of this waste by methods outlined in Section 5.5 should be encouraged.

8. INFORMATION DATABASE

An extensive amount of information is generated during the design, construction, operation and shutdown of a large nuclear facility. Some of these data could be extremely valuable to assist the decommissioning operators to plan and execute the decommissioning of the facility. Unfortunately, a large percentage of the information is not pertinent to decommissioning. The task, therefore, is to select the information required for decommissioning, and find a suitable, cost effective method of storing it so that it is available when required. This task should be completed as soon as possible after the decision for facility shutdown.

A site specific database might or might not be on a computer program, depending on the size of the facility. The database should provide an accurate and detailed description of the facility so that the most effective ways of removing equipment and structures can be determined and should include items such as:

— Material specification and analyses
— Construction prints and drawings of the plant as it was built and lists of these items
— Photographs taken during construction and installation
— Data from the national database, if any, which are particularly relevant to the particular facility
— Information and drawings of specialized procedures and equipment used during maintenance which could be of value during decommissioning
— Unusual occurrences during operation which might affect decommissioning procedures
— Radiation and contamination maps
— Scale models.

The construction prints, drawings and photographs could be stored on microfilm or other process.

A site specific database for a particular facility could save money and time during decommissioning; however, creating and maintaining such a database for decommissioning purposes alone would probably not be cost effective. The database could also be used during the construction and operational phases and then be downgraded to eliminate data not relevant to decommissioning and stored intact for long periods of time, perhaps for a long as 100 years.

It is imperative that at least two separate physical locations be established to maintain the documentation over these long periods of time to ensure that the information is not lost through events such as fires.

If computer databanks are used to store information for long periods of time, the systems associated with the databank will require frequent updating to ensure compatibility with current systems. Data retrieval procedures will also require updating. Computer databanks should be backed up by hard copies of the data.

Generic data outlining techniques or experience from previous decommissioning or decontamination can be stored on a national or international database which can be computer accessed such as the one available at the US Remedial Action Program Information Center (RAPIC) at Oak Ridge [85]. The RAPIC is a central clearinghouse for information concerning scientific, technical, regulatory and socio-economic aspects of radioactive sites requiring remedial action or decommissioning. The RAPIC uses their computerized database to provide specialized services such as annotated bibliographies [13] and literature searches for governmental, industrial, academic and foreign requests.

One type of generic data which are not currently in such databases and could eventually be of assistance in planning and costing future decommissioning operations is the inclusion of unit cost factors. These factors are used to describe the decommissioning process by breaking it down into elementary activities such as removal of pumps and cutting pipes. A computerized cost model to use such factors has been proposed [86].

9. CONCLUSIONS AND RECOMMENDATIONS

The information presented in this report confirms the general consensus amongst technical experts that sufficient experience has been gained so far to demonstrate that decommissioning can be carried out without unacceptable impact on man or his environment. However, additional development work on decommissioning engineering, equipment and tools is necessary to improve the economics and further reduce personnel exposure.

It is recommended that:

(1) the development of 'exempt quantity or concentration' criteria be done to permit the segregation of 'inactive' waste for unrestricted release or reuse;

(2) further work be done on the development of assay instrumentation for:

    (a) the segregation of active from inactive decommissioning wastes including on-line systems where necessary

    (b) the characterization of radioactive decommissioning wastes, especially those containing fissile materials particularly in concentrations above levels acceptable for disposal in low and intermediate level waste repositories;

(3) the IAEA give more emphasis to the decommissioning of non-reactor nuclear facilities;

(4) Member States review existing databases for currently operating facilities to determine if adequate information for decommissioning is being stored and if it can be retrieved later in a suitable manner without great difficulty;

(5) a more detailed analysis of the means of reducing occupational exposures during decommissioning be done;

(6) an assessment be made of the technology, safety and economics of recycling radioactive materials from decommissioning;

(7) continued development of remote handling and robot technology specifically aimed at decommissioning tasks should be encouraged. Further work on the development of automated vehicles, power transmission,

adaptive control, man–machine interface, sensor technology, viewing systems, artificial intelligence, radiation tolerant design, communications, etc. is required. These developments will in part be advanced in non-nuclear fields, but adaptation for decommissioning requirements will be needed. Such equipment can be used to replace workers in highly active environments to take radiation readings, clean up areas, place shielding blocks, hold dismantling tools, etc.;

(8) when adequate data are available, a new table on "Quality Assurance Programme Documentation: Decommissioning" should be added to Annex A in Ref. [87] to go along with similar tables on construction, commissioning and operation.

No conclusions are drawn at this time on the best way to finance the decommissioning of a particular facility. The method selected will depend on the type of facility, the national policy and many other country and facility specific parameters. The sections presented in this report on costing–financing are only to give general guidance on what methods are available.

In estimating the costs to decommission a particular facility, caution must be used in doing the estimates if disposal costs are not well known and if no similar facility has been decommissioned. Considerable contingency should be provided to cover these and other uncertainties.

Annex A

INVENTORY OF NUCLEAR FACILITIES OF INTEREST TO DECOMMISSIONING

Annex A gives a summary of some of the nuclear facilities of interest to decommissioning and is compiled from a variety of references such as [9] and [88–95]. The recent detailed review by the NEA(OECD) provided a valuable and updated review of facilities in the OECD countries [88]. This annex is not a complete summary since there are numerous small facilities, such as some of those listed in Ref. [89], which have not been included. In addition, data on certain countries are not readily available.

Table IV lists abbreviations used in Tables V–VII. Tables V–VII list decommissioning projects of power reactors, research and test reactors, and other facilities, respectively.

The distinction between prototype power reactors and test reactors is not always well defined. For convenience, the two nuclear ships and the Hanford and SRP plutonium production reactors are listed with the power reactors.

TABLE IV. ABBREVIATIONS USED IN TABLES V–VII

(1) **Countries**

BEL	– Belgium		NL	– Netherlands
CAN	– Canada		NOR	– Norway
CZE	– Czechoslovakia		POL	– Poland
FRA	– France		SPA	– Spain
FRG	– Federal Republic of Germany		SWE	– Sweden
IND	– India		SWI	– Switzerland
ITA	– Italy		UK	– United Kingdom
JAP	– Japan		USA	– United States of America

(2) **Organizations, facilities and reactor types**

AGR	Advanced Gas Cooled Reactor
AHCF	Aqueous Homogeneous Critical Facility
ALRR	Ames Laboratory Research Reactor
ANL	Argonne National Laboratory (USA)
ARE	Aircraft Research Reactor
ASTR	Aerospace Test Reactor
BNFL	British Nuclear Fuels Limited
BONUS	Boiling Nuclear Superheater Power Station
BORAX	Boiling Water Reactor Experiment
BWR	Boiling Water Reactor
CEA	Commissariat à l'Energie Atomique (France)
CEC	Commission of the European Communities
CVTR	Carolina Virginia Tube Reactor
DFR	Dounreay Fast Reactor
DMTR	Dounreay Materials Test Reactor
EBR-1	Experimental Fast Breeder Reactor
EBWR	Experimental Boiling Water Reactor
FBR	Fast Breeder Reactor
FBTR	Fast Breeder Test Reactor
GCR	Gas Cooled, Graphite Moderated Reactor
GMR	Graphite Moderated Reactor, Water Cooled
GTR	Ground Test Reactor
HNPF	Hallam Nuclear Power Facility
HTGR	High Temperature Gas Cooled, Graphite Moderated Reactor
HTR	Hitachi Training Reactor
HWCTR	Heavy Water Components Test Reactor
HWGCR	Heavy Water Moderated, Gas Cooled Reactor
HWLWR	Heavy Water Moderated, Boiling Light Water Cooled Reactor
HWR	Heavy Water Reactor
INEL	Idaho National Engineering Laboratory (USA)
IRL	Industrial Reactor Laboratories
JAERI	Japan Atomic Energy Research Institute
JPDR	Japan Power Demonstration Reactor

TABLE IV   (cont.)

JRR-1	Japan Research Reactor
KKN	Kernkraftwerk Niederaichbach Reactor
KRBA	Kernkraftwerk RWE/BAG Gundremmingen
KSTR	Kema Suspension Test Reactor
KWL	Kernkraftwerk Lingen Reactor
LGR	Light Water Cooled, Graphite Moderated Reactor
LMFBR	Liquid Metal Fast Breeder Reactor
LWBR	Light Water Breeder Reactor
LWCF	Light Water Critical Facility
LWCHW	Light Water Cooled, Heavy Water Moderated and Cooled Reactor
LWGR	Light Water Cooled, Graphite Moderated Reactor
MCF	Mitsubishi Critical Facility
MSRE	Molten Salt Reactor Experiment
MTR	Materials Test Reactor
MZFR	Mehrzweckforschungsreaktor
NEA	Nuclear Energy Agency (OECD)
OCF	Ozenji Critical Facility
OECD	Organization for Economic Co-operation and Development
OMCDR	Organic Moderated and Cooled Demonstration Reactor
OMRE	Organic Moderated Reactor Experiment
ORNL	Oak Ridge National Laboratory (USA)
PHWR	Pressurized Heavy Water Moderated and Cooled Reactor
PM	Portable Medium Power Plant
PNPF	Piqua Nuclear Power Facility
PTR	Pool Type Reactor
PWR	Pressurized Water Reactor
SCF	Sumitomo Critical Facility
SCGMR	Sodium Cooled, Graphite Moderated Reactor
SEFOR	Southwest Experimental Fast Oxide Reactor
SGHWR	Steam Generating Heavy Water Reactor
SNEF	Saxton Nuclear Experimental Facility
SPERT	Special Power Excursion Test
SRE	Sodium Reactor Experiment
SRP	Savannah River Plant
UKAEA	United Kingdom Atomic Energy Authority
USDOE	United States Department of Energy
USNRC	United States Nuclear Regulatory Commission
VBWR	Vallecitos Boiling Water Reactor
WAGR	Windscale Advanced Gas Cooled Reactor
WBR	Water Boiler Reactor
WTR	Westinghouse Test Reactor

TABLE V. REVIEW OF DECOMMISSIONING PROJECTS – POWER REACTORS

| Power reactor | | | Power output (MW(e)) | Start-up | Final shut-down [a] | Current status | | Future plans | | Comments (S = Stage) | Refs |
Name	Location	Type				Stage	Completion date	Stage	Completion date		
CAN Gentilly-1	Quebec	HWLWR	250	1970	1979	SS	1986			Static state–extension of S-1	[88]
CZE NPP-A1	Jaslovske, Bohunice	HWGCR	110	1972	1979	1					[9]
FRA Chinon-A1	Avoine-Chinon	GCR	70	1963	1973	1		3		Partial S-1 before 1984	[9, 88]
G-2	Marcoule	GCR	45	1958	1980	1		3		S-2 and S-3 being planned	[9, 88]
G-3	Marcoule	GCR	45	1959	1983			3		Decommissioning starts 1993	[9, 88]
FRG KKN	Niederaichbach	HWGCR	100	1972	1974	1	1981	3	1990	Engineering in progress	[88]
KRBA	Gundremmingen	BWR	250	1966	1977	1	1987				[88]
KWL	Lingen	BWR	256	1968	1977	1				In S-1 for 20 years	[9, 88]
MZFR	Karlsruhe	PWR	58	1965	1984	1	1987	2		Started 1984	[88]
NS Otto Hahn	Nuclear Ship	PWR	15	1968	1979	3	1982				[88]
HDR	Grosswelzheim	BWR	23	1970	1971						[9]
ITA Garigliano	Caserta	BWR	160	1964	1978	1				S-1 being planned	[88]
JAP JPDR	Tokai	BWR	12.5	1963	1976			3	1990		[9, 88]
SWE Agesta	Stockholm	PHWR	12	1963	1974	1	1975	3		Study in progress	[88]

60

Country	Exp. Nuc. P.S.	Location	Type	Power (MW)		Last op.[a]				Status	Ref
SWI		Lucens	HWGCR	7	1967	1969	2	1973	3	Experimental power station	[88]
UK	DRF	Dounreay	FBR	14	1959	1977	1	In progress	3		[88]
UK	WAGR	Sellafield	AGR	33	1962	1981	1	In progress	3	S-3 plan complete	[88]
USA	BONUS	Puerto Rico	BWR	15	1964	1968	2	1970		Entombed	[9, 90]
USA	CVTR	Parr, SC	HWR	17	1962	1968	1	1968		Safe storage	[9]
USA	Dresden-1	Morris, IL	BWR	210	1960	1978	3	1974			[94]
USA	Elk River	Elk River, MN	BWR	23.8	1964	1968	1	1975			[83, 88, 89]
USA	Enrico Fermi-1	Lagoona Beach, MI	FBR	57	1963	1973	1	Complete			[9]
USA	GE EVESR	Alameda Co., CA	BWR	17	1962	1967	2	1968			[9]
USA	HNPF	Hallam, NB	SCGMR	75	1963	1964	1	1980	3	Entombed	[88, 90]
USA	Humboldt Bay	Eureka, CA	BWR	50	1962	1976	1	1971	3 (2010)	Safe storage	[93]
USA	Indian Point-1	Buchanan, NY	PWR	257	1961	1974	1	1978	3	Safe storage	[9]
USA	NS Savannah	Charleston, SC	PWR	80	1964	1971	1	1968		Safe storage	[9]
USA	Pathfinder	Sioux Falls, SD	BWR	58	1967	1967	1	1975	2		[88]
USA	Peach Bottom-1	Peach Bottom, PA	HTGR	40	1964	1974	2	1971	2028	Entombed	[88]
USA	Piqua	Piqua, OH	OMCDR	15		1966	3	1968		Entombed	[90]
USA	PM-3A	Antarctica	PWR	9			1	1975			[9]
USA	Pu prod. reactors	Aiken, SC	GMR		1953		1			2 of 5 reactors SD	[90]
USA	Pu prod. reactors	Richland, WA	PWR		1944	1971	1	1971		8 reactors, to be entombed	[64]
USA	Shippingport	Shippingport, PA	PWR	72	1955	1982			3 (1988)	Complete dismantling	[9]
USA	TMI-2	Three Mile Island	PWR	905	1978	1979				Decommissioned after accident	[9]
USA	VBWR	Almedo Co., CA	BWR	50			1				[9]

a Refers to last year of operation.

TABLE VI. REVIEW OF DECOMMISSIONING PROJECTS – RESEARCH, TEST AND MINOR POWER REACTORS

Name	Reactor Location	Type	Power output (MW(th))	Start-up	Final shut-down	Current status Stage	Completion date	Future plans Stage	Completion date	Comments (S=Stage)	Refs
CAN Zeep	Chalk River	Tank	0.1	1945	1970	1	Complete			Used as museum	[9, 88]
FRA Cesar	Cadarache	GCR		1964	1974	3	Complete				[9, 88]
EL 2	Saclay	HWR	2.8	1952	1965	2	1968				[9, 88]
EL 3	Saclay	HWR	18	1957	1979	1	1982	3	1984		[9, 88]
G1	Marcoule	GCR	46	1956	1968	2	1981				[9, 88]
Minerve	Font-aux-Roses	LWR	1	1959	1976	3	1977				[9, 88]
Nereide	Font-aux-Roses	LWR	0	1959	1982	3		3		Start decommissioning in 1986	[9, 88]
Pegase	Cadarache	LWR	35	1962	1975	3	Complete			Used for fuel storage	[9, 88]
Peggy	Cadarache	LWR	0	1961	1975	3	Complete				[9, 88]
Rapsodie	Cadarache	FBR	40	1967	1982						[88]
Triton	Font-aux-Roses	LWR	5	1959	1982			3		Start decommissioning in 1986	[88]
Zoe	Font-aux-Roses	HWR	0	1948	1975	2	Complete				[9, 88]
FRG FR-2	Leopoldshafen	HWR	44	1960	1981	2	1984				[9]
FRN	Neuherberg	LWR	1	1972	1982	3	1983			Hall used for air pollution experiments	[9]
IND Zerlina	Trombay	HWR	0	1961	1982	3					[9]
ITA Avogadro	Saluggia, TO.	PTR	7	1959	1972	1	1975			Used for spent fuel storage	[9, 88]
Galilei	Pisa	PTR	5							Decommissioning option being evaluated	[88]
Ispra-1	Ispra, VA.	Tank	5	1958	1974	1	1975				[9]
Rana	Casaccia, Rome	PTR	0	1965	1981			3	1985		[9, 88]
RB1	Bologna	U enriched graphite	0	1960	1981			3	1985		[9, 88]
RB2	Bologna	Graphite H$_2$O	0.1	1963	1981			3		Under evaluation	[9, 88]
Ritmo	Cassacia, Rome	PTR	0	1965	1978			3	1985		[9, 88]
Rospo	Cassacia, Rome	Organic H$_2$O	0	1963	1975			3		S-3 complete	[9, 88]
JAP AHCF	Tokai	AHCF	0	1961	1967	3	1979				[9, 88]
HTR	Ozenji	PTR	0.1	1961	1975	2	1976				[9]
JRR-1	Tokai	WBR	0	1357	1968	1	1970				[9]
MCF	Omiya	LWCF	0	1969	1973	3	1975				[9]
OCF	Ozenji	LWCF	0	1962	1973	3	1975				[9]
SCF	Tokai	LWCF	0	1966	1970	3	1971				[9]

Country	Reactor	Location	Type	Power (MW)	Operation start	Operation end	Stage	Status	Stage	Status	Remarks	Ref
NL	DR	Schiphol	0	0			3	Complete	3	Complete	DR = Demonstration Reactor	[88]
	ER	Eindhoven	PWR	1			3	Complete	3	Complete	ER = Education Reactor	[88]
	KSTR	Arnhem	PWR		1974	1977	3		3	1986	Urania–Thoria suspension	[88]
NOR	Jeep	Kjeller					2		2	Complete		[88]
	Nora	Kjeller					3		3	Complete		[88]
POL	EWA	Swierk	LWR	10	1958				3		Partial decommissioning by 1988	[95]
SWE	R-1	Stockholm	HWR	1	1954	1970	3		3	1983		[91]
SWI	Diorit	Würenlingen	HWR	30	1960	1977	3		3			[9]
UK	Bepo	Harwell	GCR	6.5	1946	1968	2	Complete	3		Planned to S-3	[88]
	DMTR	Dounreay	HWR	20	1958	1969	1/2	Complete	2		To partial S-2	[88]
	Dragon	Winfrith	HTGCR	10	1964	1976	2	1970			Area used for labs	[9]
	Jason	Langley	LWR				3		3	Complete		[88]
	Merlin	Aldermaston	LWR				3		3	Complete		[88]
USA	ALRR	Ames, IA	HWR	5	1965	1977	2	1981				[88, 89]
	ARE	Oak Ridge, TN		1	1951	1954	3	1955			Fluid fuel	[9]
	ASTR	Fort Worth, TX	PTR	10	1951	1971	3	1974				[9]
	B & W	Lynchburg, VA	PTR	6	1964	1971	1	1973				[9]
	CEER	Puerto Rico	PTR	1	1961	1975	3	1973			Training reactor	[88]
	EBR-1	Scottsville, ID	FBR		1951	1963	3	Complete				[88]
	EBWR	Argonne, WI	BWR	100	1956	1967	1	Complete	3		Start S-3 in 1987	[9]
	GTR	Fort Worth, TX	HWR	10	1972	1974	3	Complete				[89]
	HWCTR	Aiken, SC	HWR		1961	1964						[9]
	IRL	Plainsboro, NJ	PTR	5	1958	1975	3	1977			Partially dismantled	[89]
	MSRE	Oak Ridge, TN	Molten Salt		1965	1969	3	1979	3	1995	Start decommissioning in 1985	[9, 90]
	OMRE	Idaho Falls, ID	OMCDR	12	1957	1963	3	1973	3		Decommissioning to S-3 planned	[9]
	Plumbrook	Sandusky, OH	SWR	100	1963	1973	1				Safe storage	[9]
	Saxton	Saxton, PA	PWR	23	1962	1972	1	1980			Auxiliary facilities remain	[89]
	SPERT	Idaho Falls, ID	LWR				3	1984			Building reused	[90, 92]
	SRE	Chatsworth, CA	SCGMR	20	1957	1965	3				Homogeneous fuel	[9]
	W. Reed	Washington		50	1962	1971	3	Complete				[9]
	Westinghouse TR	Waltz Mills, PA	Tank	60	1959	1962	1				Test reactor; safe storage	[9]

TABLE VII. REVIEW OF DECOMMISSIONING PROJECTS – OTHER MAJOR FACILITIES

| Facility | | Operation | | Current status | | Future plans | | Comments | Refs |
Description/name	Location	Start-up	Final shut-down	Stage	Completion date	Stage	Completion date	(S = Stage)	
BEL Radium factory	Mol			3	Complete				
CAN Westinghouse APD lab.	Hamilton							Natural uranium fuels	[9]
FRA AT-1 fuel reprocessing facility	La Hague	1969	1979	1	Start 1982	3	1990	Fast reactor fuel	[9, 88]
Attila fuel reprocessing facility	Font.-aux-Roses	1966	1975	3	Complete	3	In progress	Pilot reprocessing plant	[9, 88]
BT-18 Pu, metallurgy facility	Font.-aux-Roses		1981	3	Complete				[9]
Elan IIA	Saclay	1968	1970			3		Cs & Sr source fabrication	[9]
Elan IIB	La Hague	1970	1973			3		8100 GBq Cs-137	[9]
Guégnon processing plant	Guégnon		1980	1	Start 1981	3		Ore treatment	[9, 88]
Gulliver vitrification plant	Marcoule	1965	1967	1	Start 1981	3		Pilot plant	[9]
Ore treatment plant	Le Bouchet					3			[9]
FRG Alkem fuel fabrication facility	Karlsruhe	1965	1972	1	Start 1971	3	Complete	(U, PuO2) pilot fab. plant	[88]
IND Fuel reprocessing plant	Trombay	1965				3		Partially decommissioned to S-3	[9]
NL MOX fuel pilot plant	Petten			3	Complete				[88]
UO2 fuel pilot plant	Petten			3	Complete				[88]
NOR U reprocessing plant	Kjeller			2	Complete			Research plant	[88]

	Facility	Location	Start	End	Status	No.	Year	Comments	Ref.
SPA	Pu laboratory	Madrid			Complete	3			[88]
	Uranium oxide plant	Madrid			Complete	3			[88]
	Uranium plant	Madrid			Complete	3			[88]
SWE	AB Atomenergi U facility	Stockholm			Complete	3			[88]
	National defense Pu lab.	Stockholm			Complete	3			[88]
UK	BNFL decontamination centre	Sellafield	1950s	1982		3	1983	Pu & F.P. activity	[88]
	BNFL gaseous diffusion plant	Capehurst			In progress	1		Plan to reuse buildings	[88]
	Electromagnetic sep. & labs	Harwell				3		Lab. now decommissioned to S-3	[88]
	FBR fuel reprocessing plant	Dounreay	1960	1971	1975	2		Cells reused	[9]
	Fuel storage/decanning pond	Sellafield	1950s	1960		1		Still used for storage	[88]
	Thorium factory	Widnes	1940s	1976	1978	3			
USA	ANL-E surplus facilities	Argonne, WI	1942		In progress	1	1986	6 radioactive facilities	[88]
	ANL-Pu fuel fabrication facility	Argonne, WI	1959	1974	1982	3		16 GB lines & equipment	[88]
	GE advanced fuels lab.	Vallecitos, CA	1962	1979	1982	3		GB lines – Pu fuels	[90]
	Hanford 100 area facility	Richland, WA	1971		In progress	1		Support for production reactors	[88, 89]
	Hanford 200 area facility	Richland, WA			In progress	1		56 surplus fac. & Pu plant	[88, 89]
	INEL surplus facility	Idaho Falls, ID	1949		In progress	1		Many facilities	[88, 89]
	Kerr McGee LWR fuel fabrication	Crescent, OK				1			[9]
	Mound surplus facility	Miamisburg, OH	1960s	1977	In progress	1	1994	Pu processing plant	[88, 89]
	Nuclear fuel services Pu fuels	Erwin, TN				3		Prep. for safe storage	[9]
	Nuclear material deviation facility	Santa Susana, CA	1967	1980	In progress	1	1983	Building to be reused	[90]
	ORNL surplus facility	Oak Ridge, TN	1943		In progress	3		76 active labs & facilities	[88, 89]
	Westinghouse Pu fuel lab.	Harmour, PA	1966	1979		3	1982	Advanced fuel fabrication	[88, 89]

Annex B

DISASSEMBLY TECHNIQUES

B.1. INTRODUCTION

The dismantling of power reactors and other fuel cycle facilities generally involves the segmentation of reactor vessels, vessel internals, tanks, piping and other components. In most facilities to be decommissioned, demolition or surface scarification of some concrete will also be necessary to remove activated or contaminated areas of the structures. Annex B summarizes the special and conventional tooling and techniques used in these applications. The activities and techniques described below and in Table VIII in this annex include:

(a) Segmenting activated vessels and internals

 — Arc saw
 — Plasma arc torch
 — Oxyacetylene cutting
 — Thermite reaction lance

(b) Segmenting piping, tanks and other components

 — Arc saw
 — Plasma arc torch
 — Oxyacetylene cutting
 — Thermite reaction lance
 — Explosive cutting
 — Hacksaws and guillotine saws
 — Abrasive cutters
 — Circular cutters

(c) Concrete demolition and surface decontamination

 — Controlled blasting
 — Wrecking ball or slab
 — Backhoe mounted rams
 — Flame cutting
 — Rock splitter
 — Demolition compounds
 — Wall and floor sawing
 — Core stitch drilling
 — Pavement breakers
 — Drill and spall
 — Scarifiers

The techniques that are suitable for more than one type of activity will only be described once. The reader is referred to more detailed publications such as the US DOE Decommissioning Handbook [10] for a description of the operating parameters and cutting or demolition performance of each of the pieces of equipment.

B.2. SEGMENTING ACTIVATED VESSELS AND INTERNALS

Components such as reactor vessels, vessel internals, thermal shields, and structures and supports in the vicinity of the reactor vessel will become quite radioactive owing to the neutron flux emanating from the reactor core. Therefore, the dismantling and removal of these components while the radioactive inventory is still significant must be done remotely and with adequate radiation shielding for personnel protection.

Typically, the vessel walls of 1100 MW(e) light water reactors consist of 20–33 cm thick carbon steel, with stainless steel cladding approximately 0.6 cm thick. The vessel internals are made of stainless steel and usually range in thickness up to 7 cm, although certain structural sections of some pressurized water reactors are greater in thickness. In addition to steels, aluminium is a typical material of construction for low power test reactor vessels and internals.

B.2.1. Arc saw cutting

The arc saw is a development of Retech. Inc. [96]. It is a circular, toothless saw blade that cuts any conducting metal without physical contact with the workpiece. The cutting action is obtained by maintenance of a high current electric arc between the blade and the material being cut. The blade can be made of any electrical conducting material such as tool steel, mild steel or copper with equal success.

Rotation of the blade is essential to operation but rotational speed is not a critical parameter; 300 to 1800 rev/min is acceptable. Blade rotation affects removal of the molten metal generated by the arc in the kerf of the workpiece. The molten material condenses in the form of highly oxidized pellets as it is expelled from the kerf. Rotation aids in cooling of the blade and maintenance of its structural integrity; the depth of cut is limited by blade diameter. The arc saw can cut through a 30 cm thick vessel section using a 90 cm diameter blade. The arc saw can operate under water or in air. Figure 4 is a sketch of the arc saw and one design of installation.

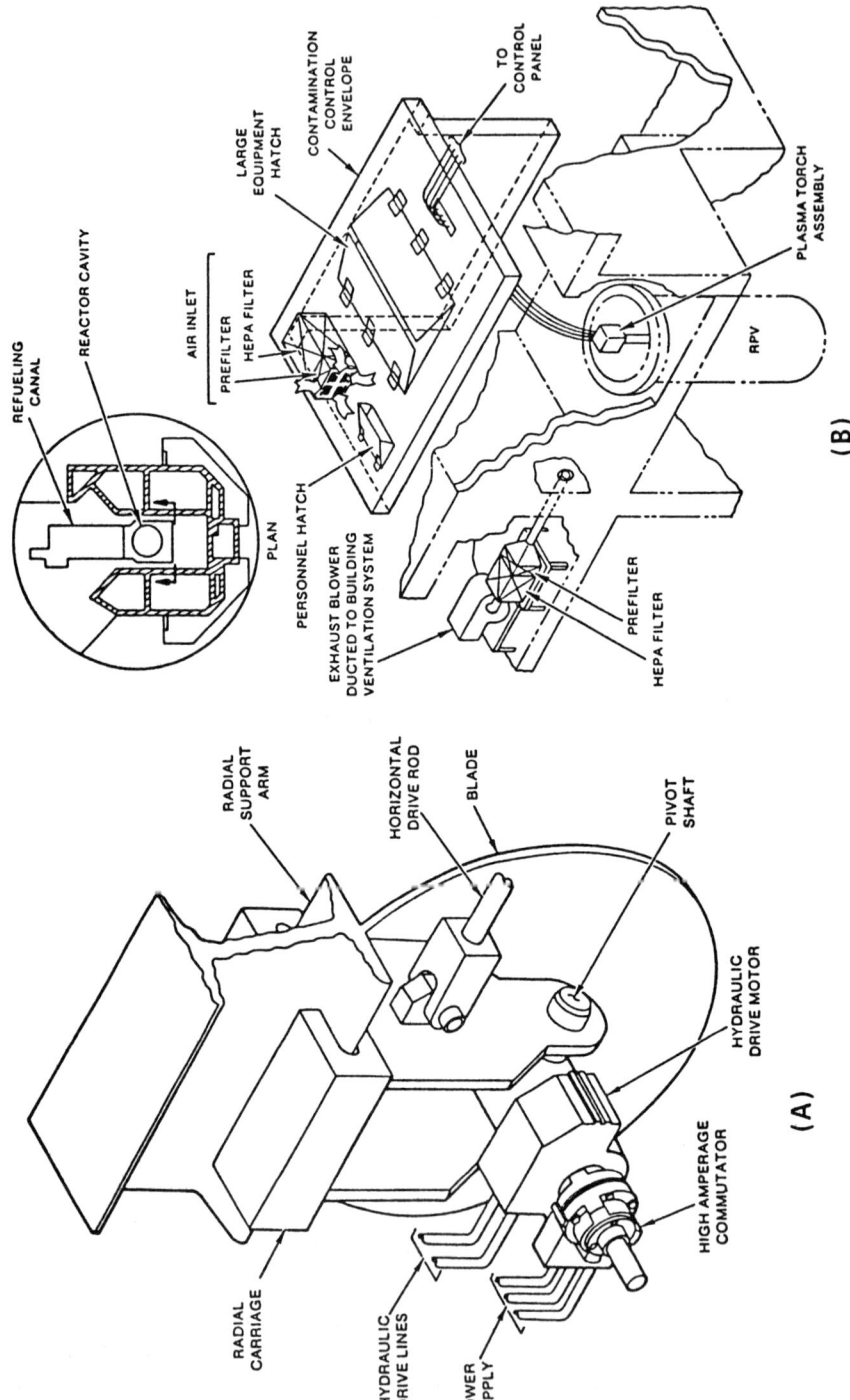

(A)

RADIAL SUPPORT ARM

HORIZONTAL DRIVE ROD

BLADE

PIVOT SHAFT

HYDRAULIC DRIVE MOTOR

HIGH AMPERAGE COMMUTATOR

RADIAL CARRIAGE

HYDRAULIC DRIVE LINES

POWER SUPPLY

(B)

REFUELING CANAL

REACTOR CAVITY

PLAN

LARGE EQUIPMENT HATCH

CONTAMINATION CONTROL ENVELOPE

TO CONTROL PANEL

PLASMA TORCH ASSEMBLY

RPV

AIR INLET

PREFILTER

HEPA FILTER

PERSONNEL HATCH

EXHAUST BLOWER DUCTED TO BUILDING VENTILATION SYSTEM

PREFILTER

HEPA FILTER

FIG. 4. Sketch of the arc saw head (A) and assembly arrangement (B).

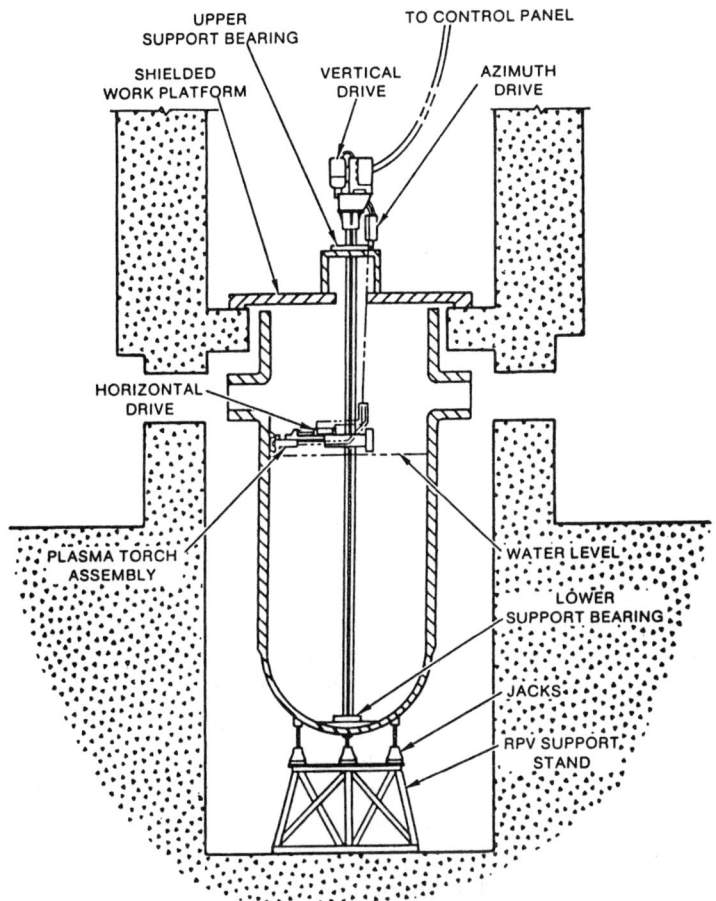

FIG. 5. Plasma torch system for reactor vessel.

B.2.2. Plasma arc torch

The plasma arc torch cuts by a direct current arc between a tungsten electrode and any conducting metal. The arc is established in a gas such as argon that flows through a constricting orifice in the torch nozzle to the workpiece.

The constricting effect of the orifice on both the gas and the arc results in very high current densities and high temperatures in the stream (10 000–24 000°C). The maximum depth of cut is approximately 17 cm.

The stream of plasma consists of positively charged ions and free electrons. The plasma is ejected from the torch nozzle at a very high velocity and, in combination with the arc, melts the contacted workpiece metal and blows the molten metal away. The torch is most effective in cuts starting at a free edge, but it is also capable of making piercing cuts anywhere on the surface.

Figure 5 shows a schematic representation of a torch assembly unit in position for segmenting a reactor vessel. This figure depicts the actual equipment used for disassembly of the Elk River reactor and sodium reactor experiment vessels [97].

Studies on the application of the plasma torch in France have shown that it is possible to cut stainless steel, up to 20 cm thickness, under water. During cutting in water, it is important to filter the water to permit better vision and avoid contamination. While cutting in air, it is important to trap at the source the dust particles and aerosols which are produced.

B.2.3. Oxyacetylene cutting

Oxyacetylene cutting, sometimes referred to as oxygen burning, consists of a flowing mixture of a fuel gas and oxygen ignited at the orifice of a torch. The fuel gas may be acetylene, Mapp gas, propane or hydrogen. A hand held torch is the general method of usage of this process, although it is readily adaptable to automated positioning and travel. The cutting tip of the torch consists of a main oxygen jet orifice surrounded by a ring of preheater jets. The fuel gas is exothermically oxidized through the preheater jets. When the metal to be cut reaches approximately 800°C the main oxygen jet is turned on and the heated metal is 'burned' away, leaving a reasonably clean cut surface.

Oxygen burning depends on the rapid exothermic oxidation of the metal to be cut. Therefore, only ferrous metals that will undergo this process can be cut with an oxygen burning torch.

An oxygen burning torch is ordinarily unable to cut stainless steel, aluminium and other non-ferrous or ferrous/high per cent alloy metals because of the formation of refractory oxides (e.g. CrO_2 and Al_2O_3) with high melting point temperatures. An iron powder or an iron/aluminium powder flowing mixture can be introduced at the torch nozzle and the torch flame temperature significantly increased. The iron/aluminium powder results in a higher temperature owing to a thermite reaction. The increased flame temperature will melt the refractory oxides formed by the oxygen.

B.2.4. Thermite reaction lance

The thermite reaction lance is an iron pipe packed with a combination of steel, aluminium and magnesium wires that maintains a flow of oxygen gas. The lance cuts are achieved by a thermite reaction at the tip of the pipe, completely consuming all constituents. Temperature at the tip range from 2250 to 5500°C, depending on the environment (in air or under water) and the ambient conditions of that environment. The lance is ignited in air by a high temperature source such as an oxygen burning torch or an electric arc. Typical lances are 3 m in length and 10 or 6 mm in diameter.

The thermite reaction lance can be used in air or under water. The operational procedure is the same in either environment except that the lance must always be ignited in air and the incident angle relative to an underwater workpiece must be considered in order to preserve the operator's visibility since many bubbles form during the process. The rate for metal cutting has been reported as generating approximately a 2.5 cm diameter hole at the rate of 30 cm depth per minute, provided the molten metal is free to flow away from the kerf.

B.3. SEGMENTING PIPING, TANKS AND COMPONENTS

The removal of piping, tanks and ancillary components is a major activity in any dismantling programme, particularly when radioactive materials are present. Removal of these radioactive components must be accomplished in a controlled manner to contain radioactivity and prevent the spread of contamination. Removal may be done manually if no significant direct radiation hazards are present or if local shielding can be used effectively. Remote removal may be necessary when highly contaminated or activated systems preclude direct worker access. In the latter case remote cutting would be required to accomplish the segmenting activity.

Typically, reactor system piping in a light water reactor consists of (1) carbon steel piping with diameters up to 200 cm and wall thicknesses up to 15 cm and (2) stainless steel piping generally of smaller diameter and wall thickness. Tanks are fabricated of either stainless steel or carbon steel. Tank diameters vary from about 1 to 15 m, with wall thicknesses commensurate with the tank's pressure rating. Ancillary components such as pipe hangers and supporting beams are fabricated largely of carbon steel. The segmenting processes suitable for cutting these components are: arc saw, plasma arc torch, oxyacetylene and thermite reaction lances, which are described in Section B.2, as well as explosive cutting, hacksaws, guillotine saws and abrasive circular cutters, which are described below.

A summary of the application characteristics of each process is shown in Table VIII.

B.3.1. Explosive cutting

An explosive cutter consists of an explosive core such as cyclotrimethylene-trinitramine (RDX) or pentaerythritol tetranitrate (PETN) surrounded by a casing of lead, aluminium, copper or silver. Cutting is accomplished by a high explosive jet of detonation products of combustion and deformed casing metal. The jet forms a directed shock wave that severs the target material.

The cutter is chevron shaped with the apex pointing away from the material to be cut. When detonated, the explosive core generates a shock wave that fractures

TABLE VIII. APPLICATION CHARACTERISTICS FOR SEGMENTING PROCESSES

Method	Application	Use	Relative cost	Notes[a]
Spark erosion	All metals up to 5 cm	R	High	
Arc saw	All metals up to 90 cm	R	High	
Plasma arc torch	All metals up to 15 cm	P, R, S	High	Cost for a system to cut 10 cm steel is about US $25 000
Oxyacetylene cutting	Mild steel, all thicknesses	P, R, S	Low	A hand held torch costs about US $3000
Thermite reaction lance	All metals, all thicknesses	P	Low	
Explosive cutting	All metals up to 15 cm	R	High/very high	
Hacksaws and guillotine saws	(a) Piping up to 45 cm dia.	P, R	Low	Prices range from US $2000 to 4000 depending on size and type
	(b) Piping or stock up to 60 cm dia. All metals	S	Low	
Abrasive cutter	(a) Piping or stock up to 5 cm chord. All metals	P	Low	Cost range US $400 to 800. Abrasive wheels cost about $5
	(b) Piping or stock up to 20 cm chord. All metals	S	Low	
Circular cutter	Piping up to 15 cm dia, with wall thickness to 7.5 cm. All metals	P, R	Low	Base cost about US $7000

[a] Costs are in US $ and are very approximate to show the order of magnitude.

P = Portable application where personnel bring the process equipment to the components being disassembled.

R = Remote application where remotely operated mechanisms are required to segment components.

S = Stationary application while material is brought to a permanently established work station for segmenting.

1. CHARGE PLACEMENT

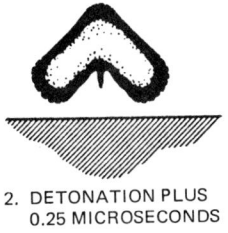

2. DETONATION PLUS
0.25 MICROSECONDS

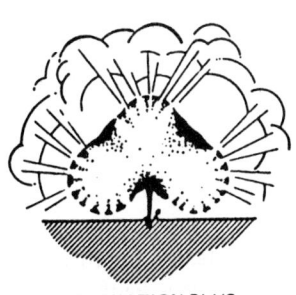

3. DETONATION PLUS
0.50 MICROSECONDS
CRACK INITIATION

4. DETONATION PLUS
1.0 MICROSECONDS
CRACK PENETRATION

FIG. 6. Detonation sequence (Source: Explosive Technology).

the casing inside the chevron and propels the casing into the material to be cut. Figure 6 shows the detonation sequence during a cut.

Explosive cutting is normally used either when the geometry of an object being cut is too complex to employ other methods or when several cuts must be made simultaneously. It is ideally suited for cutting concentric pipes.

B.3.2. Hacksaws and guillotine saws

Hacksaws and guillotine saws are relatively common industrial tools used for cutting all metals with a reciprocating action, hardened steel saw blade. These saws use mechanical methods for segmentation rather than the previously discussed thermal methods. This offers two distinct advantages: fire hazards are reduced and radioactive contamination control is simpler because there are no fumes or gases.

73

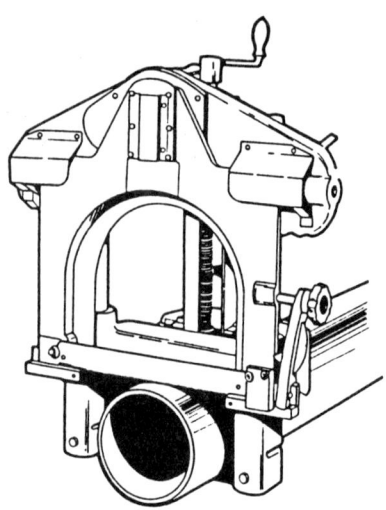

FIG. 7. *Portable air powered guillotine saw.*

Hacksaws and guillotine saws are the tools frequently selected for cutting piping systems because of their low operating cost, high cutting speed and ease of contamination control. They can be applied in either portable or stationary modes.

Portable power hacksaws are clamped with a chain to a pipe in a position such that the blade contacts the underside of the pipe. This allows the weight of the motor to advance the blade into the workpiece around the chain mounted pivot point. An operator may increase the feed pressure manually by applying downward force on the motor body or by suspending weights from the body. In general, blade lubrication is not necessary.

A portable guillotine saw also clamps by chain to a pipe but the saw and motor are mounted above the cut allowing the weight of the unit to advance the saw into the workpiece. In general, blade lubrication is not necessary. Figure 7 shows a typical air powered guillotine saw. Motors for either type portable saw may use air or electricity for motive power.

Portable power hacksaws can cut piping up to 36 cm in diameter. Cutting time varies with the material being cut, use of lubricant if any, and the force applied to the blade. As a general rule, a 20 cm diameter Schedule 40 pipe can be cut in about 8 minutes by a power hacksaw.

B.3.3. Abrasive cutters

An abrasive cutter is an electrically, hydraulically or pneumatically powered wheel formed of resin bonded particles of aluminium oxide or silicon carbide.

FIG. 8. Circular cutting machine.

Usually the wheel is reinforced with fibreglass matting for strength. It cuts through the workpiece by grinding the metal away, leaving a clear kerf.

The cutting process generates a continuous stream of sparks making it unsuitable for use near combustible materials. Contamination control is a significant problem since the swarf particles are removed in very small pieces. Cutters may be fitted with a swarf containment system that acts to limit the spread of contamination. Water lubricants also tend to limit the spread of contamination.

B.3.4. Circular cutters

A circular cutting machine is a self-propelled circular saw that cuts as it moves around the outside circumference of a pipe on a track. The machine may be powered either pneumatically, hydraulically or electrically and is held to the outside of the pipe or component by a guide chain that is sized to fit the outside diameter. A guide ring is available if very precise cuts are necessary.

The cutter blades are made of hardened steel and their number may be varied to change the thickness of the cut. Wall thicknesses of up to 7 cm may be cut on pipes with outside diameters up to 6 m. Figure 8 shows a typical circular cutting machine [98]. The maximum cutting depth in carbon steel is limited to 2 cm per pass. Multiple passes are necessary for thicker pipe wall thicknesses.

Contamination control is maintained by vacuuming the chips from the cut if required. Cutting lubricants are applied in a fog spray and cause almost no dispersal or contamination. Since the cutting is by mechanical methods there is no fire hazard.

B.4. CONCRETE DEMOLITION AND SURFACE DECONTAMINATION

Nearly every decommissioning operation requires the demolition or surface decontamination of concrete structures. Activated concrete in the region immediately surrounding the core beltline represents the most difficult concrete removal activity. This is due to the relatively high radiation fields and potential for release of radioactive particulates during demolition. During operation leaks of radioactive fluid may have contaminated floor or wall surfaces of a facility; these surfaces are resistant to non-destructive cleaning methods because of the porosity of concrete. Although non-radioactive concrete structures do not represent any unique demolition difficulty, the volume of such concrete coupled with significant reinforcement represents a difficult dismantling task.

Typically, the biological shield surrounding a reactor vessel or the walls of a hot cell consists of massive sections (1 to 3 m thick) of standard or high density concrete. In most cases the biological shield may be heavily reinforced to meet seismic design criteria.

If the facility is to be converted to other uses, it may be advantageous to remove the contamination by removing a layer without demolishing the structures, particularly in the case of thick walls (greater than 0.6 m).

B.4.1. Controlled blasting

Controlled blasting is generally recommended for demoliton of massive or heavily reinforced thick concrete sections. The process consists of drilling holes in the concrete, loading them with explosives and detonating using a 1 to 3 millisecond delayed firing technique. The delayed firing increases fragmentation and controls the direction of material movement. Delayed firing also reduces the vibration impact on adjacent structures. Each borehole fractures radially during the detonation. The radial fractures in adjacent boreholes form a fracture plane. The detonation wave separates the fractured surfaces and moves the material towards the structure's free face. Figure 9 illustrates a typical blasting round for massive concrete demolition and explains the terminology used in designing a blast; for example, the burden is the distance from the free face.

The selection of the best type of explosive requires an evaluation of the properties of the explosive and concrete itself. A blasting expert should be consulted to select the best explosive for the purpose.

A blasting mat (composed, for example, of automobile tire sidewalls tied together) is placed over the blast area. Continuous fog sprays of water should be used before, during and after the blast to hold down dust. The exposed reinforcing bar may then be cut with an oxyacetylene torch or bolt cutter.

Controlled blasting was used to demolish the 2.8 m thick steel reinforced radioactive biological shield of the Elk River reactor.

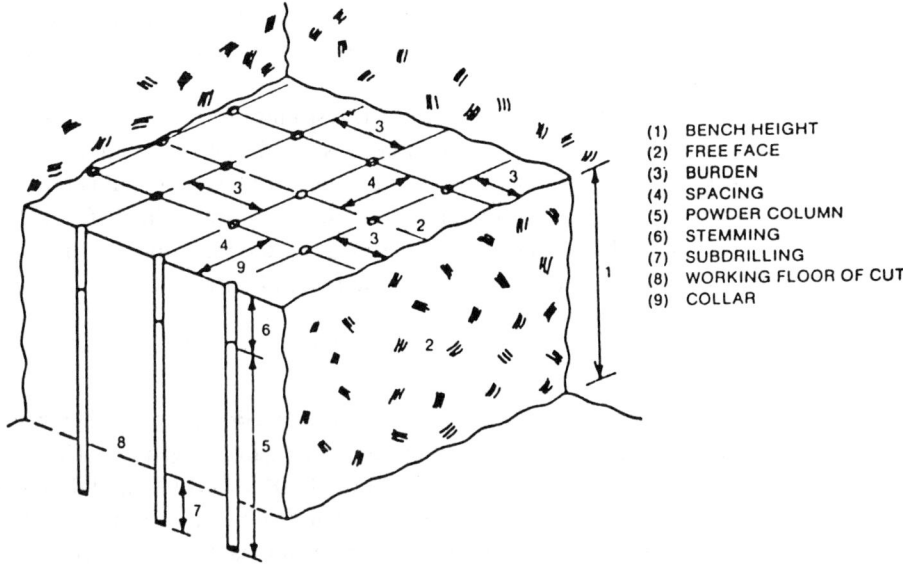

(1)	BENCH HEIGHT
(2)	FREE FACE
(3)	BURDEN
(4)	SPACING
(5)	POWDER COLUMN
(6)	STEMMING
(7)	SUBDRILLING
(8)	WORKING FLOOR OF CUT
(9)	COLLAR

FIG. 9. Blasting round.

B.4.2. Wrecking ball or slab

The wrecking ball is used for demolition on non-reinforced or lightly reinforced concrete structures less than 1 m thick. The equipment consists of a 2 to 5 Mg ball or flat slab suspended from a crane boom. The ball may be used in either of two techniques to demolish structures. The preferred method is to raise the ball with a crane 3 to 6 m above the structure and release the cable brake allowing the ball to drop onto the target surface. The maximum height of structure is limited to about 30 m. A 5 Mg ball would require a 200 Mg crane for the maximum height. This method develops good fragmentation of the structure with maximum control of the ball after impact. The second method is to swing the ball into the structure using a suck line for recovery after impact. The maximum heigt of structure is limited to about 15 m because of the crane instability during the swing and after impact. The latter method is not recommended because the target area is more difficult to hit and the ball may ricochet off the target and damage adjacent structures while putting side loads on the crane boom. The flat slab may only be used in the vertical drop mode but offers the advantage of being able to shear through steel reinforcing rods as well as concrete.

B.4.3. Backhoe mounted rams

Backhoe mounted rams are used for concrete structures less than 0.6 m thick with light reinforcement. The method is ideally suited for low noise, low vibration demolition and for interior demolition in confined areas. The equipment consists of an air or hydraulic operated impact ram with special or chisel points mounted on a backhoe arm.

The ram starts impacting as soon as there is resistance to the point and stops when breakthrough occurs or when the ram head is lifted. The ram delivers about 600 blows per minute at up to 2000 joules of energy per blow, depending on the size of the ram head. Many sizes of air and hydraulic rams are available from several suppliers. With the ram head mounted on a backhoe the operator has about a 7 m reach and the ability to position the ram in limited access structures.

B.4.4. Flame cutting

Flame cutting of concrete consists of a thermite reaction process whereby a powdered mixture of iron and aluminium oxidizes in a pure oxygen jet. The temperatures in the jet are approximately 2000 to 5000°C, causing rapid decomposition of the concrete in contact with the jet. The mass flow rate through the flame cutting nozzle clears away the decomposed concrete and leaves a clean kerf. Reinforcing rods in the concrete add iron to the reaction to sustain the flame and assist the reaction.

B.4.5. Rock splitter

The rock splitter fractures concrete by hydraulically expanding a wedge into a predrilled hole until tensile stresses are large enough to cause fracture. The tool consists of a hydraulic cylinder that drives a wedge shaped plug between two expandable guides (called feathers) inserted in the predrilled hole. The rock splitter unit is powered by a hydraulic supply system and operates at 50 MPa pressure. The hydraulic unit may be powered by air pressure, gasoline engine or electric motor sources.

Units are available to develop splitting forces approaching 3.2 MN. The maximum lateral expansion of the feathers is approximately 2 cm. Concrete may be separated at a fracture line using a backhoe mounted ram or similar equipment. The reinforcing rod in reinforced concrete must be cut before separation is possible. Additional holes and fractures will be necessary to expose the reinforcing rod for heavily reinforced concrete.

The splitter is ideal for fracturing concrete in limited access areas where large air rams cannot operate. The process is relatively quiet except for hole drilling and is used extensively for demolition near hospitals and densely populated areas.

B.4.6. Concrete demolition compounds

At least two types of concrete demolition compounds are available, BRISTAR [99] and S-MITE [100]. They are expanding compounds that are poured into predrilled holes and cause tensile fractures in the concrete upon hardening. BRISTAR is a proprietary compound of limestone, siliceous material, gypsum and slag. The powdered compound is mixed with water and kneaded to a fluid paste. The paste is filled into holes drilled in a fracture line of predetermined burden, spacing and depth. No hole caps are required if the hole depth is at least 6 times the hole diameter. Pressure will develop to over 30 MPa within 20 hours.

Since concrete tensile strengths range up to about 4 MPa, low grade concretes are likely to fracture easily. Cracks will form and propagate along the fracture line. The crack width will be about 5 cm after 15 hours. The fractured burden may then be removed with a pavement breaker, backhoe or bucket loader. If reinforcing rod is encountered it must be cut separately. The compound is not classified as a hazardous substance and can be readily stored and handled. There is no noise or vibration (except for drilling holes), or flyrock, dust or gas release. Contamination control is only required during drilling and removal.

B.4.7. Wall and floor sawing

Wall and floor sawing is generally used when disturbance of the surrounding material must be kept to a minimum. A diamond or carbide wheel is used to abrasively cut a kerf through the concrete. The blades can cut through reinforcing rods although the rods tend to break off the blade diamonds. The blade is rotated by an air or hydraulic motor. For most applications the saw will be mounted on a guide that also supports the saw's weight. The operator manually advances the blade into the work. The dust produced by the abrasive cutting is controlled using a water spray. The abrasive blade produces no vibration, shock, smoke, sparks or slag and is relatively quiet. Figure 10 shows a photograph of a large diameter wall saw developed by the CEGB in the United Kingdom [101].

Thicknesses up to one metre have been cut with concrete saws. The maximum thickness of cut is approximately equal to one-third of the blade diameter. The saw cuts approximately 0.2 m² per minute of cut surface regardless of thickness [102]. Cutting can be done either manually or remotely, depending on the size of the saw.

B.4.8. Core stitch drilling

Core stitch drilling consists of close pitched drilling of holes in concrete using a diamond or carbide tipped drill bit in an electric or fluid driven rotary drill. This method is not recommended for reinforced concrete because the remaining reinforcing rods inhibit shearing. The centre lines of the holes are located to correspond to the desired breaking plane in the concrete. The hole pitch is such that there is very little concrete left between the adjoining holes (less than 1/2 the radius of the holes). When a line of holes has been drilled along the breaking plane, bars are inserted into the holes and force is applied to the free end of the bars in a

FIG. 10. *A large circular saw developed to cut through reinforced concrete.*

line perpendicular to the breaking plane to shear the remaining concrete. Alternatively, a wrecking ball may be dropped onto the piece to be removed to shear the remaining concrete.

B.4.9. Pavement breakers

Pavement breakers remove concrete and asphalt by mechanically fracturing localized sections of the surface. Fracturing is caused by the impact of a hardened tool steel bit such as a chisel point. The bit is driven in a reciprocating motion by either a compressed air or hydraulic fluid pressure source. These breakers (also called jackhammer and /or pneumatic drills) have an approximate mass of 15 to 45 kg and are intended for use on floors. The breakers deliver about 1500 blows per minute at up to 130 joules of energy per blow depending on the size of the unit.

Pavement breakers are recommended for use on floors to remove small areas that are inaccessible for heavy equipment. They may also be used to expose reinforcing rods after controlled blasting to permit cutting of the rods. The chisel point may be used to scarify surface areas of concrete floors where contamination may have penetrated several inches deep in localized areas. The spread of contamination may be controlled using water or fog sprays.

Large hydraulic breakers can also be mounted on large excavators capable of working in hazardous environments under control of an operator located several hundred metres from the vehicle.

B.4.10. Drill and spall

The drill and spall technique was developed for the removal of contaminated surfaces of concrete without demolishing the entire structure. The technique consists of drilling 2 to 3 cm diameter holes about 8 cm deep and inserting a hydraulically operated spalling tool. The spalling tool bit is an expandable tube of the same diameter as a hole. The average surface removal rate is 7 m^2/h.

A tapered mandrel is hydraulically forced into the hole to spread the fingers and spall off the concrete. The holes are drilled on approximately 30 cm centres such that the spalled area from each hole overlaps the next. Figure 11 shows the spaller system in use.

The drill and spall technique is recommended for removing surface contamination that penetrates up to 5 cm into the surface. Removal of the surface radioactivity in this manner eliminates the need to dispose of large quantities of nonradioactive concrete as with other volume removal techniques. Contamination is reduced while drilling by using a filtered vacuum system. Fog sprays may be used to wet the surface and reduce contamination and dust levels.

FIG. 11. Drill and spall rig.

B.4.11. Scarifiers

The scarifier technique is best suited for the removal of thin layers (up to 2.5 cm in thickness) of contaminated concrete. The tool, marketed under the trade name of Scabbler [103], consists of pneumatically operated piston heads that strike the surface to chip off the concrete. The piston heads are available in either 5-point or 9-point tungsten carbide bit sizes depending on the degree of surface roughness allowable. The 5-point bit has 0.6 cm high points and the 9-point bit has 0.3 cm high points.

The pistons are mounted in a wheeled floor chassis that is available in 5, 7 and 9 piston sizes. The chassis is pushed along the floor to remove the surface layer. The chassis can be modified to include a HEPA filtered vacuum exhaust system to capture contaminated dust. Other tool models include a 3-piston wall Scabbler that may be spring counterbalanced to relieve the tool weight. Smaller hand held units are available but are not intended for large surface area removal. Figure 12 shows the Scabbler floor tool and typical 5-point and 9-point bits.

FLOOR MODEL

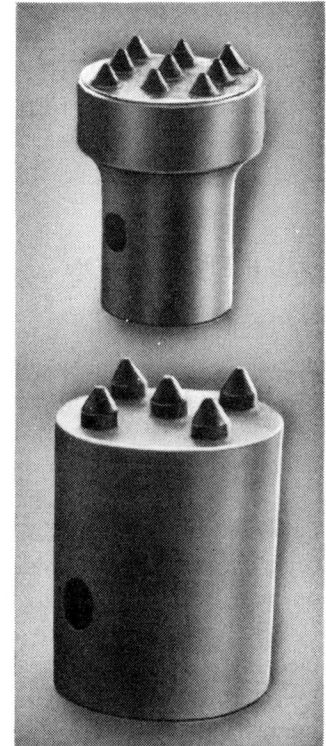

SCABBLER BITS

FIG. 12. Floor model Scabbler and bits.

The Scabbler tool is recommended for applications where the concrete surface is to be used after decontamination. The scarified surface is generally level with a coarse finish resulting from the 9-point bit. The coarse surface is suitable for bonding to a concrete finish cap and the smoother surface suitable for epoxy, polymer and similar finishes.

The concrete surface removal rate is about 4 m²/h per bit for the floor Scabbler, representing 28 m² per hour for a 7-piston unit. The 3-piston Scabbler will remove about 7 to 11 m² of surface per hour. The tungsten carbide tool bits have an average working life of 80 hours under normal use.

Annex C

REMOTELY CONTROLLED EQUIPMENT
FOR DECOMMISSIONING

C.1. INTRODUCTION

Different types of automation and robots have been used with success for some decades for industrial handling assembly and manipulation.

In the nuclear industry a wide range of specialized manipulators and equipment has been, and is now being, developed to perform remote tasks such as inspection, maintenance, repair and refurbishment.

The use of such automatic and/or remotely controlled devices is one of the important methods of reducing man exposure to radiation and contamination occurring during the decommissioning operations at nuclear facilities. As a consequence it may also decrease decommissioning costs.

In Annex C the terms 'robot' and 'manipulator' will have the following meaning:

A **robot** is a programmable handling machine that has a memory, can be trained and can be retaught easily when changed to a new job. This latter capability is the characteristic difference between robots and other pieces of automated equipment, although the flexibility of numerically controlled equipment is also high. Robots consist of mechanical components, actuators, controls and sensors, and generally have many degrees of freedom.

A **manipulator** has many of the features of a robot, but is usually operated directly under some form of manual control, which may be remote. Programmed control of a manipulator can be accomplished (producing a form of robot), just as manual control of a robot is possible through an appropriate control system.

Both robots and manipulators can range from very simple, uncomplicated mechanisms, with only a few degrees of freedom, to the more complex electronic force reflecting master–slave manipulators where the ratio of force or displacement experienced by the operator can be changed.

For decommissioning work the following components are important for both robot and manipulator applications:

— Task analysis
— Remote control technology
— Advanced mechanical engineering
— Simulation technology
— Remote sensing equipment
— Man–machine interface

The majority of commercially available robots and manipulators are listed in directories which detail their performance and applications [104, 105].

Annex C

(a) surveys industrial robots and selected control techniques which should have potential application to decommissioning
(b) surveys the use of basic and sophisticated manipulators and remotely controlled equipment in the nuclear industry, especially in decommissioning
(c) provides an overview of criteria for designing decommissioning robots and remote manipulators.

C.2. INDUSTRIAL ROBOTS

Most robots have an articulated mechanical arm on to which can be attached various end effectors such as a gripper, grinder, paint sprayer, welding gun or pneumatic wrench. The design and operation of these end effectors which, for the robot arm, take the place of the hand on the human arm, are as important as the design of the robot arm itself. The kinematic functions of a robot (or a manipulator) and end effectors are usually evaluated by comparison with the original multipurpose manipulator, the human arm and hand. A person can, without conscious programming, instantaneously will his arms and hands to follow any desired path in space by continuous path control.

In addition to this spatial control, the arm and wrist can reorient an object through three planes of rotation giving a total of six degrees of freedom (three translational axes and three rotational axes) as shown in Fig. 13. Furthermore, the human end effector (the hand) has 22 separate movements. When combined with the human sensory functions (vision, hearing, touch, force, and spatial location), the feedback system and the ability of the human brain to automatically select the sensory feedback most appropriate to every movement, the human arm is a very versatile and complex machanism. It is also very strong (a strength to weight ratio of about 5) and lightweight.

To simulate the spatial and reorientation functions of the human arm and wrist, a robot arm or manipulator must have at least six degrees of freedom. Often as many as 8 or 9 degrees of freedom are used to permit the robot arm to reach around obstacles. In most cases the 22 movements of the human hand are replaced with a simple pincer device, although in very special cases anthropomorphic hands with articulated and powered fingers have been developed [106]. The actuators of a robot can be operated pneumatically, hydraulically, electrically, mechanically or in some combination of the four basic drives.

The most difficult task in developing a robot arm is to simulate the control functions of the human arm. The development of specialized computers, processors and memories has ushered in a new phase of robot control. The designer can increase the intelligence of the robot by using mathematical equations for complex motions and more complicated sensory devices.

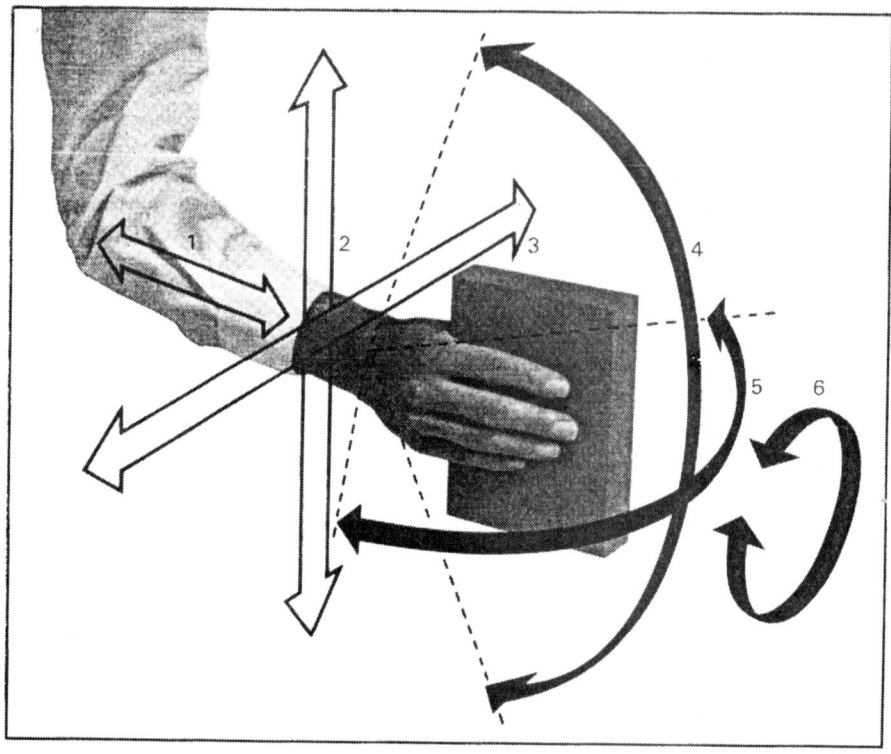

FIG. 13. The 6 degrees of freedom of the human arm – 3 translational axes and 3 rotational axes.

The design of the robots and control systems currently in use vary widely and range from simple, limited sequence robots to sophisticated computer controlled robots.

Many thousands of limited sequence robots, mainly of the 'pick and place' type are being used successfully in factories throughout the world. These robots only require a limited number of sequential actions to do their job and do not require the control of motion of more sophisticated robots. Each step of the sequence of operations is preprogrammed and controlled by an electric or pneumatic signal from a plugboard control panel. Mechanical stops are generally used to limit the motion of each joint. Although most devices have no ability to get feedback from their working environment, some have been combined with other devices to permit some intelligence. The technology associated with limited sequence robots would not appear to have much application in decommissioning tasks since the robots are relatively difficult to reprogramme.

In the more sophisticated commercial robots (Fig. 14) one of the most important components is the control system which dictates when and how the robot

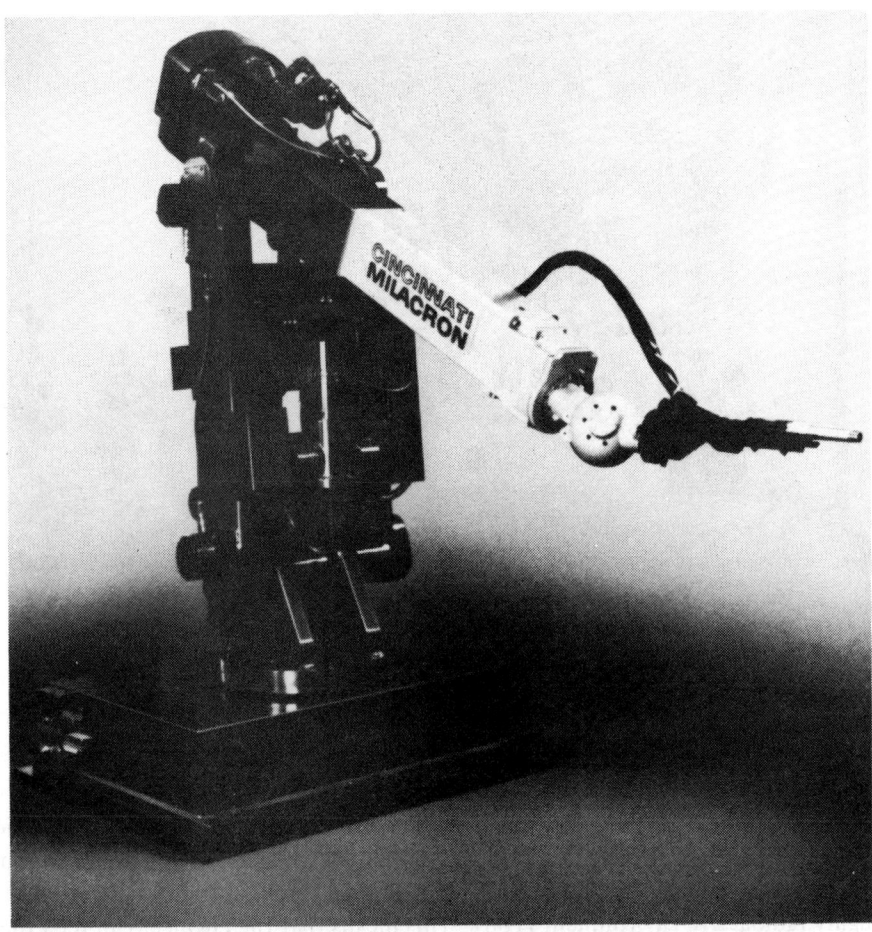

FIG. 14. *A sophisticated computer controlled robot* [107] *which has great versatility and load carrying ability.*

hardware will perform its tasks. The control system, which consists of a computer, a high capacity memory and sensory feedback, must have simple and fast on- and off-line programming abilities and be capable of interacting with an operator and with feedback from as many sensors as are required to do the job. Sensors permit a measurement of the real time state of the system and can provide feedback to the controls. Laser range finders, television systems, tactile, force, torque and prox- imity sensors are typical. The computers store input data and complex programs for the robot, compare the measured and desired performance, generate complex sequences of desired outputs and permit communication between a human and the complex robot.

FIG. 15. Through wall master–slave manipulators (AERE Harwell photograph).

Just as important as the robot hardware and controls is the new software
technology which is being developed. For example, interactive computer graphics
systems could permit the automation engineer to put the robot operated equipment
through its paces on a computer screen rather than through 'trial and error' in a
highly radioactive environment [108]. This means that the engineer can feed the
design and location of the piping, equipment, etc., and the surrounding facility into
the graphics computer, and preprogramme the robot and its working tool before
installation. For example, if a series of pipes of various sizes and shapes have to be
cut, a robot mounted on a mobile remotely controlled vehicle, a crane or a gantry,
and having a cutting torch as an end effector could be preprogrammed to cut
through the pipes in the right sequence. Collision avoidance should be achieved
by using sensors or the software model of the environment derived from the
graphics system. This type of technology would appear to be well worth
developing for decommissioning and for maintenance using either sophisticated
robots or for more conventional kinds of remote system technology combined
with various tools.

Advanced robot systems, which are being used in a wide range of factories
for welding, foundry applications, spray painting, delicate assembly jobs, etc., will
relieve many workers of dangerous and unpleasant tasks. Direct application of
specific industrial robot systems to decontamination and decommissioning tasks in

the nuclear industry depends on activity and contamination levels but application of the technology, components and specific robots is feasible and desirable for many tasks. It must be kept in mind that robots with force, touch and visual sensory capability are available but are still under development; even modest maintenance tasks challenge current robot technology. Robot systems work best when the co-ordinates of the physical environment surrounding them are well defined and within the robot's memory. Continued investment in artificial intelligence research and development and the establishment of interface standards common to all robot suppliers will increase the flexibility of future robot applications.

In evaluating the financial attractiveness of applying robot technology to any activity, conventional return on investment and payback calculation methods should not be applied. In addition to the standard savings of the salaries of workers displaced by the robot system, other savings include the cost of items such as health insurance, less supervision, no parking space, less training, heat, light and power, no pension, sick leave or man-rem costs. A cost–benefit analysis technique has been established to allow comparison of the cost of robot applications to nuclear facility inspection [109].

C.3. REMOTE OPERATION AND MANIPULATION IN THE NUCLEAR INDUSTRY

Remotely operated equipment has been used in the nuclear industry since its inception to perform a large variety of tasks in environments which are hostile to man primarily because of high radiation fields. For example, remotely operated equipment has been used for handling, inspection, dismantling, assembly, repair, replacement and fabrication tasks in reactors, shielded cell facilities, underwater bays, reprocessing plants, fuel fabrication and radioisotope production facilities, etc.

Of most interest from a decommissioning viewpoint are the remotely operated manipulators, robots (stationary and mobile), visual and sensor technology and the computer hardware and software associated with the equipment.

The types of manipulators in use include relatively simple master–slave manipulators (Fig. 15), power manipulators (Fig. 16), sophisticated bilateral force reflecting electric manipulators in which the master and slave can be connected by direct wire, radio or laser beam [111] and the most sophisticated and dextrous computer aided master–slave servomanipulator [112, 113]. In addition, industrial manipulators (Fig. 17) could be equipped with environmentally conditioned and shielded cab enclosures and mounted on vehicles if necessary.

Automated guided vehicle systems have been used in industry for many years for a variety of tasks. Some are free ranging with optical or radio controlled guidance systems, while others follow underfloor guidance wires. Specialized track

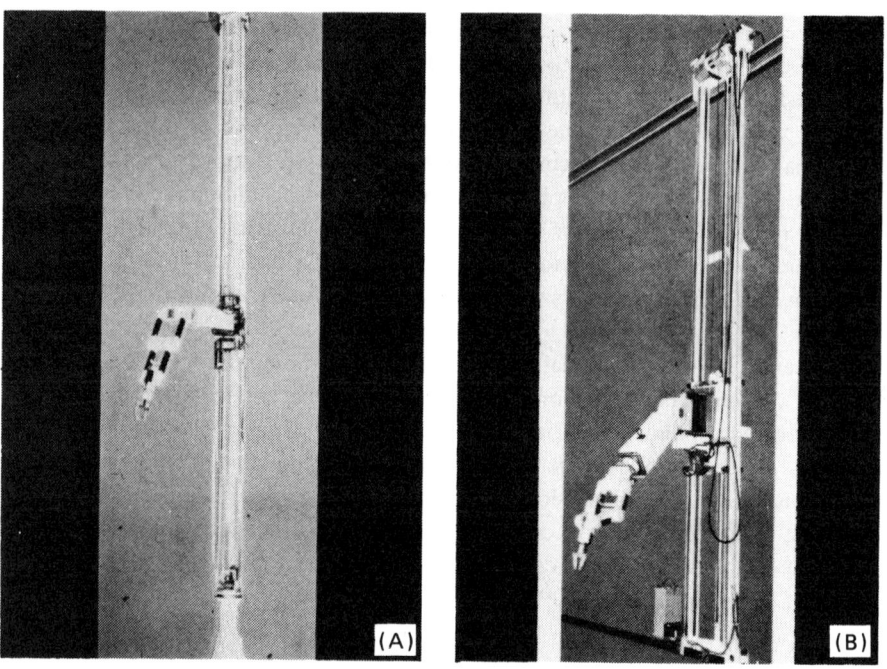

FIG. 16. Types of single unit pedestal (A), wall (B) and crane (C) mounted manipulators [110].

FIG. 17. *Heavy duty industrial manipulator having good dexterity. Could be equipped with a shielded cab* [114].

and wheel vehicles such as those shown in Fig. 18 have been developed for the nuclear industry [115–117]. For decommissioning such vehicles may be used as mobile bases for carrying manipulator arms and equipment to do work in areas with high radiation fields. A study detailing the options for using such vehicles for decommissioning has been completed for the CEC [117].

These general purpose mobile robots can advantageously replace man for multiple tasks such as surveying and monitoring. Physical measurements such as radiation levels, temperature, humidity and vibration can also be made. Simple decommissioning tasks such as scabbling and decontaminating walls and floors are possible. The load capacity of the vehicle borne manipulators will determine the extent to which small compacts can be disassembled and other tasks such as building shielding walls can be achieved.

The technique can be extended to the larger vehicles such as bulldozers, backhoes and excavators required for mass concrete demolition. Radio control systems which operate the control levers on these types of equipment are commercially available [116].

Force feedback manipulators mounted on submersibles are available for underwater use [118]. Manipulators for underwater nuclear decommissioning applications having similar characteristics could be envisaged, but would need a significant development.

FIG. 18.   One design of track vehicle used in the nuclear industry [115].

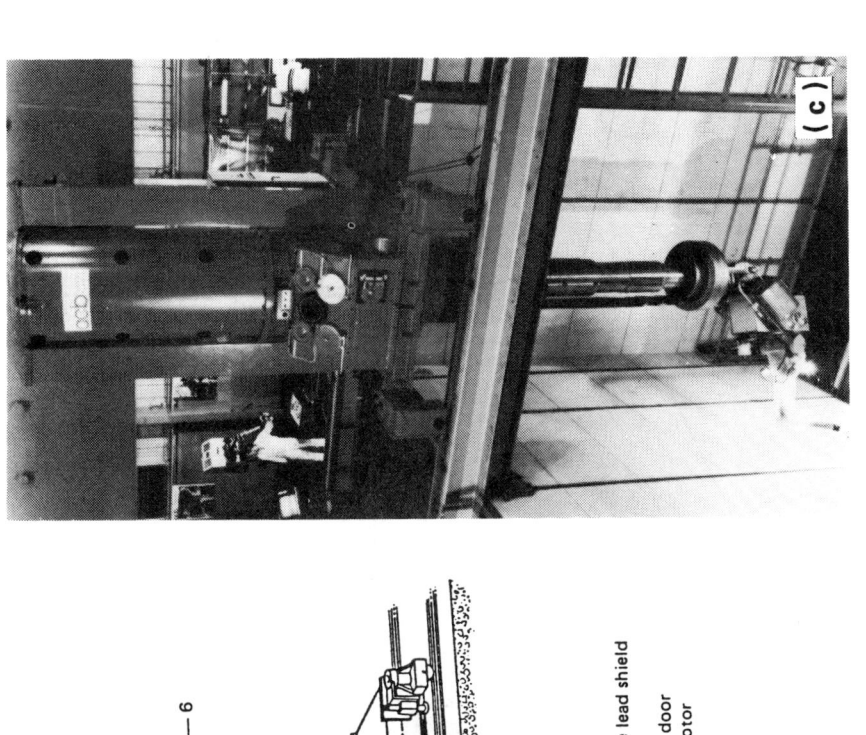

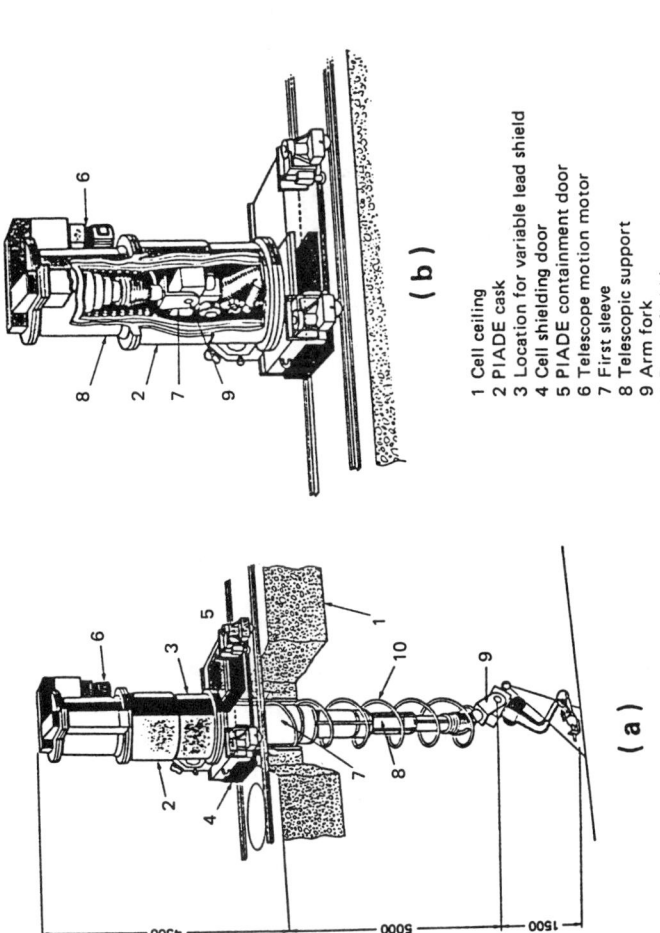

1 Cell ceiling
2 PIADE cask
3 Location for variable lead shield
4 Cell shielding door
5 PIADE containment door
6 Telescope motion motor
7 First sleeve
8 Telescopic support
9 Arm fork
10 Power fluid hoses

FIG. 19. A sophisticated servomanipulator (MA-23) equipped with television and telescopic supports with computer control [113].

(a) Extended position; (b) stowed position; (c) photograph of the servomanipulator.

93

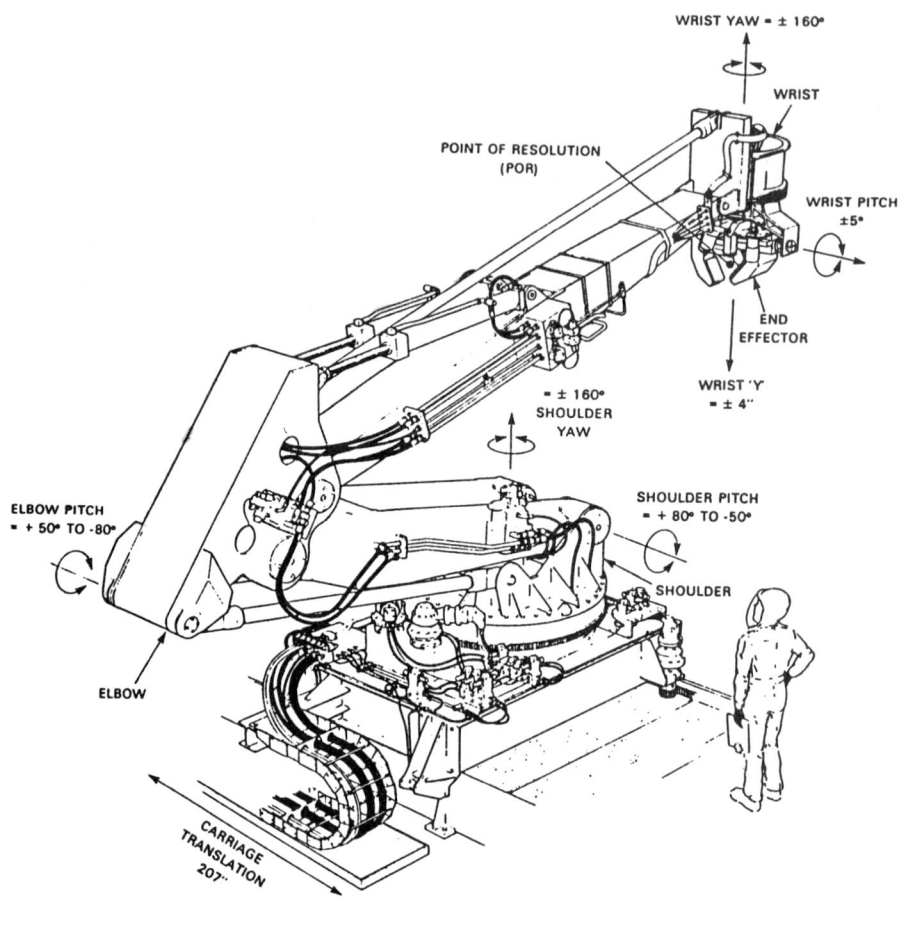

WRIST YAW = ± 160°

WRIST

POINT OF RESOLUTION
(POR)

WRIST PITCH
±5°

END
EFFECTOR

WRIST 'Y
= ± 4"

= ± 160°
SHOULDER
YAW

SHOULDER PITCH
= + 80° TO -50°

SHOULDER

ELBOW PITCH
= + 50° TO -80°

ELBOW

CARRIAGE
TRANSLATION
207"

THE RMS MECHANISM

FIG. 20. Mechanical arrangement of the remote manipulator subsystem developed to assist in the retubing of Pickering reactors in Canada.

A sophisticated maintenance servomanipulator (MA-23) with computer aided control and television cameras has been developed in France and will be used for remote maintenance and decommissioning tasks [113]. This combination of options permits the arm to be operated as a dextrous manually operated master–slave manipulator or as a computer controlled robot. Figure 19 shows the MA-23 manipulator mounted on the bottom of the telescopic boom of the PIADE self-contained telescopic linear positioner. The vertical telescope mast bearing the manipulator can be folded and brought up inside the containment flask which moves on rails.

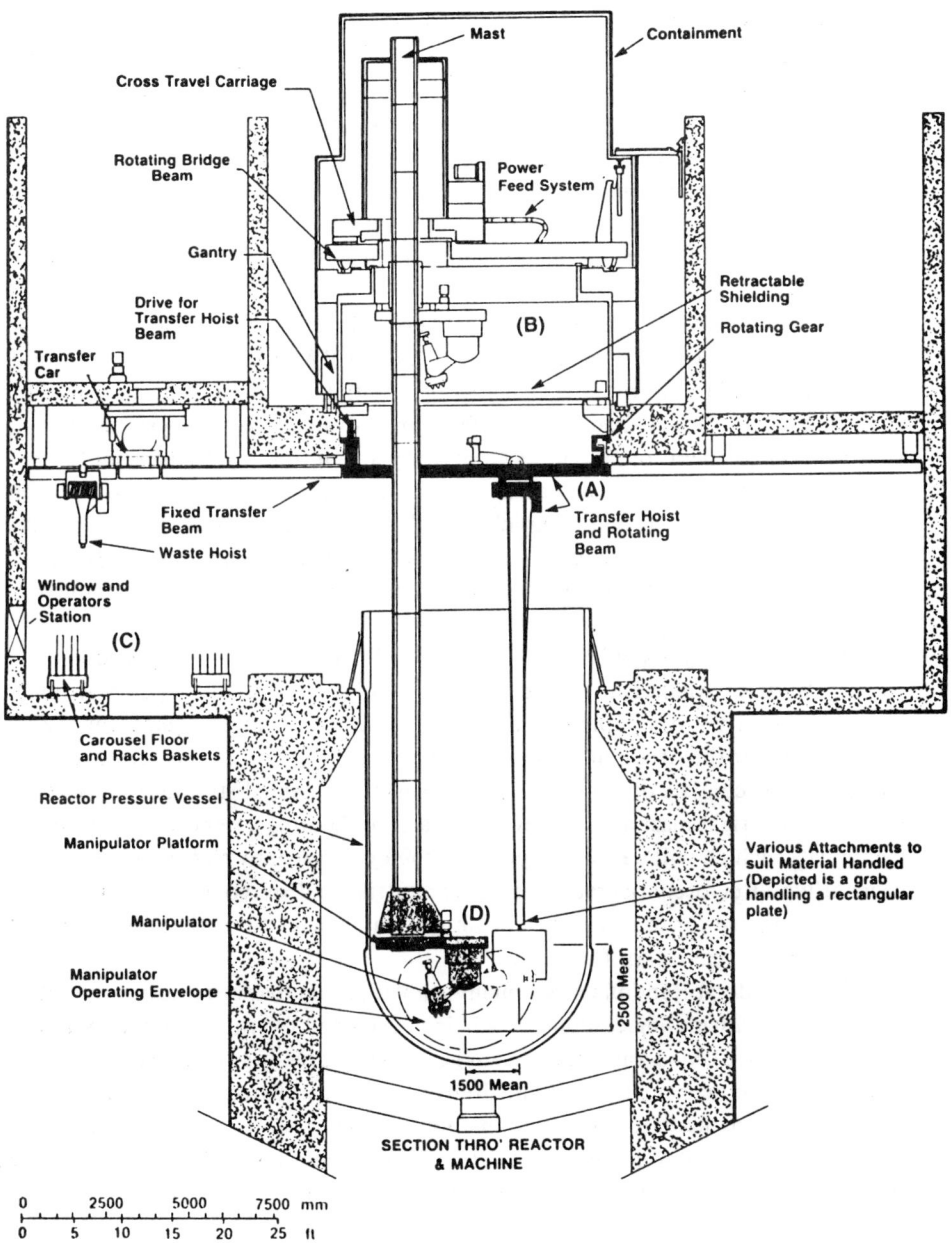

Mast

Containment

Cross Travel Carriage

Rotating Bridge
Beam

Power
Feed System

Gantry

Retractable
Shielding

Drive for
Transfer Hoist
Beam

Rotating Gear

(B)

Transfer
Car

(A)

Fixed Transfer
Beam

Transfer Hoist
and Rotating
Beam

Waste Hoist

Window and
Operators
Station

(C)

Carousel Floor
and Racks Baskets

Reactor Pressure Vessel

Manipulator Platform

Various Attachments to
suit Material Handled
(Depicted is a grab
handling a rectangular
plate)

(D)

Manipulator

2500 Mean

Manipulator
Operating Envelope

1500 Mean

SECTION THRO' REACTOR
& MACHINE

| 0 | 2500 | 5000 | 7500 mm |
| 0 | 5 | 10 | 15 | 20 | 25 ft |

FIG. 21. *Proposed decommissioning machine and transfer hoist arrangements for the decommissioning of the Windscale AGR.*

In Canada a sophisticated remote manipulator subsystem (RMS) is being developed by SPAR Aerospace for Ontario Hydro for possible use in the retubing of Pickering reactors (Fig. 20) using technology developed by SPAR for the arm used on the US space shuttle vehicles [119]. The RMS, which is one part of a co-ordinated remote manipulator and control system, would be used for a variety of handling, inspection, support and transport activities, and for manoeuvring containers in the fuelling machine vault. Atomic Energy of Canada Limited developed a complex remotely controlled arm with a viewing system and remote welding machine to repair leaking pipes located in a vault below the Douglas Point nuclear reactor [120].

A comprehensive programme is in progress in Japan to develop a robot remote handling system for the decommissioning of the 90 MW(th) power demonstration reactor JPDR [121]. During the first part of the study emphasis is being placed on the development of man–machine interfaces and advanced control systems to improve ease of operation, flexibility, dexterity and autonomy to the remote handling operations. Items such as semi-automatic trajectory control and collision avoidance for the vehicle and manipulator will be developed. Based on this work, light and heavy duty remote handling systems equipped with underwater power manipulators capable of handling 10 and 100 kg, respectively, will be available by the end of 1986.

For the decommissioning of the 33 MW(e) Windscale advanced gas cooled reactor (WAGR), which is to be completed by 1994, the United Kingdom Atomic Energy Authority is developing sophisticated remotely operated equipment [122]. The concept for the decommissioning machine system (Fig. 21) was recently completed. It consists of a movable machine gantry shielded structure supporting an extendable rigid mast and an elevating platform holding a remotely operated manipulator which will be used to perform the dextrous activities associated with dismantling. Remotely operated plasma arc cutting and stereo TV systems are being investigated for use with the decommissioning machine in the dismantling of the reactor and removal of the waste.

In the Federal Republic of Germany similar automated or remotely operated dismantling and handling equipment is being developed for the decommissioning of the 100 MW(e) Niederaichbach (KKN) gas cooled pressure tube reactor [123]. The reactor will be dismantled, segmented and packaged by remote operation using a rotary manipulator, a cutting manipulator and a crane manipulator.

The principles developed at WAGR and KKN could be applied to the design of manipulators for use in the decommissioning of other pressure vessels.

C.4. CRITERIA FOR REMOTE OPERATIONS EQUIPMENT

In the case of the nuclear industry the use of remotely operated equipment implies that it will be operated in a hostile environment in which it may get contaminated. Also man access to the equipment for adjustments, repairs and/or replacement is often difficult or hazardous and cannot be done until the unit has been decontaminated.

In designing or selecting remotely operated equipment for use in decommissioning tasks, the performance of the equipment should be carefully considered. Wherever possible, the following criteria should be followed:

— Performance of the machine even under fault conditions must be fail safe
— The machine must be reliable, durable, and able to complete the appointed task
— Often the access available for the machine will be restricted. It must be sufficient for the design to allow the deployment, operation and retrieval of the machine
— The design should cater for easy maintenance, repair and dismantling of the machine
— When used in contaminated conditions, the machine must be designed to allow easy decontamination
— The life-cycle cost should be calculated to include decontamination and/or disposal of the machine
— The man–machine interface must be considered in the control of the machine, to ensure that the performance of the operator is not degraded
— The machine should be extensively tested in mock-up facilities before it is committed to the actual task
— Connectors and fasteners should be designed for operation under remote conditions when necessay
— Components should be radiation tolerant; controls and as many other components as possible should be designed to be separate and remain outside the active zone
— The operational environmental conditions (such as temperature, pressure, humidity, dust, chemical activity) should be considered in the design
— Where practical, available and proven industrial components and/or units, should be used, even if minor modification is required
— The design should be as simple as possible, but still achieve the foregoing criteria.

REFERENCES

[1] BRAINBRIDGE, G.R., et al., Decommissioning of nuclear facilities, At. Energy Rev. 12 (1974) 145.

[2] INTERNATIONAL ATOMIC ENERGY AGENCY, Decommissioning of Nuclear Facilities, IAEA-TECDOC-179, IAEA, Vienna (1975).

[3] INTERNATIONAL ATOMIC ENERGY AGENCY, Decommissioning of Nuclear Facilities, IAEA-TECDOC-205, IAEA, Vienna (1977).

[4] Nuclear Power and its Fuel Cycle (Proc. Int. Conf. Salzburg, 1977), Vol. 4, IAEA, Vienna (1977).

[5] Decommissioning of Nuclear Facilities (Proc. Int. Symp. Vienna, 1978), IAEA, Vienna (1979).

[6] INTERNATIONAL ATOMIC ENERGY AGENCY, Manual on Decontamination of Surfaces, Safety Series No. 48, IAEA, Vienna (1979).

[7] INTERNATIONAL ATOMIC ENERGY AGENCY, Factors Relevant to the Decommissioning of Land-Based Reactor Plants, Safety Series No. 52, IAEA, Vienna (1980).

[8] INTERNATIONAL ATOMIC ENERGY AGENCY, Decontamination of Operational Nuclear Power Plants, IAEA-TECDOC-248, IAEA, Vienna (1981).

[9] INTERNATIONAL ATOMIC ENERGY AGENCY, Decommissioning of Nuclear Facilities: Decontamination, Disassembly and Waste Management, Technical Reports Series No. 230, IAEA, Vienna (1983).

[10] MANION, W.J., LAGUARDIA, T.S., Decommissioning Handbook, US Department of Energy Rep. DOE/EV-10128-1 (1980).

[11] 1982 International Decommissioning Symposium (Proc. Int. Symp. Seattle, 1982), CONF-821005, UNC Nuclear Industries, Richland, WA, Technical Information Center, USDOE (1982).

[12] Decommissioning of Nuclear Power Plants (Proc. Conf. Luxembourg, 1984), CEC, Brussels, Rep. EUR-9474, Graham & Trotman Ltd Publishers, London (1984).

[13] OWEN, P.T., et al., Nuclear Facility Decommissioning and Site Remedial Actions — A Selected Bibliography, USDOE/Oak Ridge National Lab., TN, Rep. ORNL/EIS-154/V5 (1984).

[14] INTERNATIONAL ATOMIC ENERGY AGENCY, Decontamination of Nuclear Facilities to Permit Operation, Inspection, Maintenance, Modification or Plant Decommissioning, Technical Reports Series (to be published).

[15] NUCLEAR ENERGY AGENCY (OECD), Decontamination Methods as Related to Decommissioning of Nuclear Facilities, NEA, Paris, Rep. RWM-1 (1980).

[16] BOOTHBY, R.M., WILLIAM, T.M., The Control of Cobalt Content in Reactor Grade Steels, Rep. EUR-8655, European Applied Research Report, Nucl. Sci. Technol. 5 2, Harwood Academic Publishers (1983).

[17] INTERNATIONAL ATOMIC ENERGY AGENCY, Current Practices and Options for Confinement of Uranium Mill Tailings, Technical Reports Series No. 209, IAEA, Vienna (1981).

[18] PESELLI, M., Individuazione quantitativa delle impurezze del contenitore a pressione del reattore del Garigliano, Association Euratom-ITAL, Wageningen, Rep. EUR-9167 IT (1984).

[19] MAY, S., PICCOT, D., Détermination analytique d'éléments tracés dans des échantillons de bétons utilisés dans les réacteurs nucléaires de la Communauté européenne, Commission of the European Communities, Brussels, Rep. EUR-9208 (1984).

[20] GODDARD, A.I.H., et al., Trace element assessment of low-alloy and stainless steels with reference to gamma activity, Commission of the European Communities, Brussels, Rep. EUR-9264 (1984).

[21] HEINE, W.F., Final Status Report and Safety Analysis of the Hallam Nuclear Power Facility Site and Remaining Structures, Atomics International, Canoga Park, CA, Rep. AI-AEC-MEMO-12794 (Revised) (1979).

[22] BELL, M.J., ORIGEN, the ORNL Isotope Generation and Depletion Code, Oak Ridge National Lab., TN, Rep. ORNL-4628 (1973).

[23] HONECK, H.C., ENDF/B Specifications for an Evaluated Nuclear Data File for Reactor Applications, Brookhaven National Lab., Upton, NY, Rep. BNL-50066, T-467 (1967). Also OZER, O., GARBER, D., ENDF/B Summary Documentation, Rep. ENDF/B-201 (1973).

[24] KUSNER, D.E., et al., ETOG-1, A FORTRAN IV Program to Process Data from the ENDF/B File to the MUFT, GAM and ANISN Format, Westinghouse Electric Corp., Pittsburgh, PA, Rep. SCAP-3845-1 (1969).

[25] ENGLE, W.E., Jr., A Users Manual for ANISN, A One-dimensional Discrete Ordinates Transport Code with Anisotropic Scattering, Oak Ridge Gaseous Diffusion Plant, TN, Rep. K-1693 (1967).

[26] OAK, H.D., et al., Technology, Safety and Costs of Decommissioning a Reference Boiling Water Reactor Power Station, US Nuclear Regulatory Commission Rep. NUREG/CR-0672, Vol. 2 (1980).

[27] ENGEL, R.L., et al., ISOSHLD – A Computer Code for General Purpose Isotope Shielding Analysis, Battelle Pacific Northwest Labs, Richland, WA, Rep. BNWL-236 (1966).

[28] HARBECKE, W., et al., Die Aktivierung des biologischen Schilds im stillgelegten Kernkraftwerk Lingen, Commission of the European Communities, Brussels, Rep. EUR-8801 DE (1984).

[29] EICKELPASCH, W., et al., Die Aktivierung des biologischen Schilds im stillgelegten Kernkraftwerk Gundremmingen Block A, Commission of the European Communities, Brussels, Rep. EUR-8950 DE (1984).

[30] AHLFAENGER, W., Zusammensetzung von Kontaminationsschichten und Wirksamkeit der Dekontamination, Commission of the European Communities, Brussels, Rep. EUR-9352 (1984).

[31] CONTI, E.F., Residual Radioactivity Limits for Decommissioning – Draft Report, US Nuclear Regulatory Commission, Washington, DC, Rep. NUREG-0613 (1979).

[32] INTERNATIONAL ATOMIC ENERGY AGENCY, Principles for Establishing Limits for the Release of Radioactive Materials into the Environment, Safety Series No. 45, IAEA, Vienna (1978).

[33] INTERNATIONAL ATOMIC ENERGY AGENCY, Basic Safety Standards for Radiation Protection – 1982 Edition, Safety Series No. 9, IAEA, Vienna (1982).

[34] INTERNATIONAL COMMISSION ON RADIOLOGICAL PROTECTION, Recommendations of the International Commission on Radiological Protection, ICRP, Sutton Rep. ICRP-26, Pergamon Press, Oxford (1977).

[35] INTERNATIONAL COMMISSION ON RADIOLOGICAL PROTECTION, Limits for Intakes of Radionuclides by Workers, ICRP, Sutton, Rep. ICRP-30, Pergamon Press, Oxford (1978).

[36] INTERNATIONAL ATOMIC ENERGY AGENCY, Considerations Concerning "De Minimis" Quantities of Radioactive Waste Suitable for Dumping at Sea Under a General Permit, IAEA-TECDOC-244, IAEA, Vienna (1981).

[37] INTERNATIONAL ATOMIC ENERGY AGENCY, De Minimis Concepts in Radioactive Waste Disposal - Considerations in Defining De Minimis Quantities of Solid Radioactive Waste for Uncontrolled Disposal by Incineration and Landfill, IAEA-TECDOC-282, IAEA, Vienna (1983).

[38] INTERNATIONAL ATOMIC ENERGY AGENCY, The Derivation of Exempt Quantities for Application to Terrestrial Waste Disposal, Technical Reports Series (to be published).

[39] BERVEN, B.A., et al., Radiological Survey of the Former Kellex Research Facility, Jersey City, NJ, Oak Ridge National Lab., TN, Rep. ORNL-5734 (1982).

[40] BRODZINSKI, R.L., Portable Instrumentation for Quantitatively Measuring Radioactive Contamination Levels and for Monitoring the Effectiveness of Decontamination and Decommisssioning Activities, Battelle Pacific Northwest Labs, Richland, WA, Rep. PNL-4744 (1983).

[41] BRODZINSKI, R.L., Instrumentation and Assay Procedures for Verification of the Radionuclide Content of Low-Level Waste Packages, Battelle Pacific Northwest Labs, Richland, WA, Rep. PNL-4848 (1983).

[42] SHUNK, E.R., Assay System to Measure Crate-Size Bulk Transuranic Waste, Los Alamos Scientific Lab., NM, Rep. LA-UR-83-1999 (1983).

[43] CALDWELL, J.T., Test and Evaluation of a High Sensitivity Assay System for Bulk Transuranic Waste, Los Alamos Scientific Lab., NM, Rep. LA-UR-83-2084 (1983).

[44] CRAWFORD, J.H., A Gamma Monitor for Assay of Radioactive Solide-Waste Shipments, Du Pont de Nemours (E.I.) and Co., Aiken, SC, Savannah River Plant, Rep. DPSPU-81-30-14 (1982).

[45] CSULLOG, G.W., KUPCA, S., "Waste characterization studies at CRNL", Waste Management, University of Arizona, Tucson (1984).

[46] RUOKOLA, E., A Monitoring System for Solid Wastes from Loviisa Power Station, Imatra Power Company, Helsinki, Rep. YJT-79-09 (1979).

[47] BREMNER, W.B., CLARK, M.L., SPENCE, B.W., "Operational experience relating to measurement of plutonium in solid and liquid waste streams from fast breeder reactor fuel reprocessing", Nuclear Safeguards Technology 1982 (Proc. Symp. Vienna, 1982), Vol. 2, IAEA, Vienna (1983) 463.

[48] MILLS, C.L., BIGGS, A., "Measurement of plutonium in solid wastes from fast breeder reactor fuel reprocessing using a computer controlled passive neutron counter", Safeguards and Nuclear Material Management (Proc. 5th Ann. Symp. Versailles, 1983), Joint Research Centre, Ispra (1983) 315.

[49] BREMNER, W.B., et al., An Integral Experiment Relating to Calibration and Interpretation of Plutonium Solid Waste Measurements at Dounreay Nuclear Power Development Establishment, Commission of the European Communities, Brussels, Rep. EUR-8020 EN (1982).

[50] R.J. SILLS, The management of plutonium (alpha) contaminated waste materials (PCM), Prog. Nucl. Energy 13 1 (1984) 49.

[51] Progress Reports from the Plutonium Contaminated Materials Working Party for 1982/83 (Rep. DOE/RW.83.168) and 1983/84 (Rep. DOE/RW.84.112) available from the British Lending Library, Boston Spa, Weatherby, West Yorkshire, UK.

[52] MacMAHON, T.D., Imperial College Reactor Centre, Berkshire, UK (1984).

[53] IRT CORPORATION, The ACM-110 Automated Contamination Monitor, IRT Corp. San Diego, CA.

[54] NATIONAL NUCLEAR CORPORATION, The WCM-10 Waste Curie Monitor, NNC, Mountain View, CA.

[55] INTERNATIONAL ATOMIC ENERGY AGENCY, Handling of Tritium Bearing Wastes, Technical Reports Series No. 203, IAEA, Vienna (1981).

[56] INTERNATIONAL ATOMIC ENERGY AGENCY, Conditioning of Low- and Intermediate-Level Radioactive Wastes, Technical Reports Series No. 222, IAEA, Vienna (1983).

[57] INTERNATIONAL ATOMIC ENERGY AGENCY, Treatment of Low- and Intermediate-Level Solid Radioactive Wastes, Technical Reports Series No. 223, IAEA, Vienna (1983).

[58] INTERNATIONAL ATOMIC ENERGY AGENCY, Treatment of Low- and Intermediate-Level Liquid Radioactive Wastes, Technical Reports Series No. 236, IAEA, Vienna (1984).

[59] INTERNATIONAL ATOMIC ENERGY AGENCY, Regulations for the Safe Transport of Radioactive Materials, Safety Series No. 6, IAEA, Vienna (1979).

[60] INTERNATIONAL ATOMIC ENERGY AGENCY, Disposal of Low- and Intermediate-Level Solid Radioactive Wastes in Rock Cavities — A Guidebook, Safety Series No. 59, IAEA, Vienna (1983).

[61] INTERNATIONAL ATOMIC ENERGY AGENCY, Criteria for Underground Disposal of Solid Radioactive Wastes, Safety Series No. 60, IAEA, Vienna (1983).

[62] INTERNATIONAL ATOMIC ENERGY AGENCY, Site Investigations, Design, Construction, Operation, Shutdown and Surveillance of Repositories for Low- and Intermediate-Level Radioactive Wastes in Rock Cavities, Safety Series No. 62, IAEA, Vienna (1984).

[63] INTERNATIONAL ATOMIC ENERGY AGENCY, Design, Construction, Operation, Shutdown and Surveillance of Repositories for Solid Radioactive Wastes in Shallow Ground, Safety Series No. 63, IAEA, Vienna (1984).

[64] MILLER, C.E., Jr., "Engineering and planning for the Shippingport station decommissioning project", 1982 International Decommissioning Symposium (Proc. Int. Symp. Seattle, 1982), CONF-821005, UNC Nuclear Industries, Richland, WA, Technical Information Centre, USDOE (1982).

[65] LOERCHER, G., et al., "Factors to be considered in deciding whether to decontaminate for unrestricted release", Decommissioning of Nuclear Power Plants (Proc. Conf. Luxembourg, 1984), CEC, Brussels, Rep. EUR-9474, Graham & Trotman Ltd Publishers, London (1984).

[66] LETTNIN, H.K.J. VIECENZ, H.J., "Decommissioning of the NS Otto Hahn", 1982 International Decommissioning Symposium (Proc. Int. Symp. Seattle, 1982), CONF-821005, UNC Nuclear Industries, Richland, WA, Technical Information Centre, USDOE, (1982).

[67] COMMISSION OF THE EUROPEAN COMMUNITIES, The Community's Research and Development Programme on Decommissioning of Nuclear Power Plants, Fourth Annual Progress Report (1983), CEC, Brussels, Rep. EUR-9677 EN (1985).

[68] LOERCHER, G., PIEL, W., Dekontamination von Komponenten stillgelegter Kernkraftwerke für die freie Beseitigung, Commission of the European Communities, Brussels, Rep. EUR-8704 DE (1983).

[69] INTERNATIONAL ATOMIC ENERGY AGENCY, Environmental Assessment Methodologies for Sea Dumping of Radioactive Wastes, Safety Series No. 65, IAEA, Vienna (1984).

[70] NUCLEAR ENERGY AGENCY (OECD), Decontamination Methods as Related to Decommissioning of Nuclear Facilities, An NEA Expert Group Report, NEA, Paris, Rep. RWM-1 (1980).

[71] NUCLEAR ENERGY AGENCY (OECD), Remote Handling in Nuclear Facilities, (Proc. Seminar Harwell, 1984) NEA (OECD) Paris (1984).

[72] GREGORY, A.R., CREGUT, A., "Factors to be considered in selecting a decommissioning strategy", Decommissioning of Nuclear Power Plants (Proc. Conf. Luxembourg, 1984), CEC, Brussels, Rep. EUR-9474, Graham & Trotham Ltd, Publishers, London (1984).

[73] SMITH, R.I., et al., Technology, Safety and Costs of Decommissioning a Reference Pressurized Water Reactor Power Station, Battelle Pacific Northwest Labs, Richland, WA, Rep. NUREG/CR-0130 (1978).

[74] Robert Snow Means Company, Inc., Building Construction Cost Data, Duxbury, MA, (annual editions).

[75] McGraw-Hill Information Systems Company, Dodge Guide, McGraw-Hill, New York, NY (annual editions).

[76] Bureau of Labour Statistics, Producer Prices and Price Indexes Data, US Department of Labor, Washington, DC (monthly edition).

[77] Richardson Engineering Services, Inc., Building Construction Estimating Standards, Solana Beach, CA.

[78] MANION, W.J., LAGUARDIA, T.S., An Engineering Evaluation of Nuclear Power Reactor Decommissioning Alternatives, Atomic Industrial Forum, Inc., Rep. AIF/NESP-009 (1976).

[79] GREGORY, A.R., Sizewell B Power Station Public Enquiry – CEGB, Decommissioning Addendum 2 – Comparison of Cost Estimates (1984) 24.

[80] INTERNATIONAL ATOMIC ENERGY AGENCY, Safety in Decommissioning of Research Reactors, Safety Series No. 74, IAEA, Vienna (1986).

[81] TLG ENGINEERING INC., Identification and Evaluation of Facilitation Techniques for Decommissioning of Light Water Power Reactors, Missouri Univ., Columbia, Rep. NUREG/CR-0399 (Draft-1985).

[82] MOORE, E.B., Jr., Facilitation of Decommissioning Light Water Reactors, Pacific Northwest Lab., TN, Rep. NUREG/CR-0569 (1979) prepared for USNRC.

[83] UNITED POWER ASSOCIATION, Final Elk River Reactor Program Report, Department of Energy, Chicago, IL, Rep. COO-651-93 (1974).

[84] NUCLEAR ENERGY AGENCY (OECD), Specialists Meeting on Decommissioning Requirements in the Design of Nuclear Facilities (Mar. 1980).

[85] OAK RIDGE NATIONAL LABORATORY, Remedial Action Program Information Center, A Programme sponsored by US Department of Energy.

[86] GUPTA, B., SAROUDIS, J., Methodology of a Computerized Cost Model for Decommissioning of Nuclear Power Plants, Atomic Energy of Canada Limited, Discussion paper presented at the first meeting of the Co-ordinated Research Programme on Decommissioning and Decontamination, Vienna 26-29 November 1984.

[87] INTERNATIONAL ATOMIC ENERGY AGENCY, Establishing the Quality Assurance Programme for a Nuclear Power Plant Project – A Safety Guide, Safety Series No. 50-SG-QA1, IAEA, Vienna (1984).

[88] NUCLEAR ENERGY AGENCY (OECD), Compendium of Decommissioning Activities in NEA Member Countries, NEA, Paris (1985).

[89] UNITED STATES DEPARTMENT OF ENERGY, Surplus Facilities Management Program, Program Plan (FY 1983-1987) Department of Energy, Richland, WA, Rep. RLO-SFM-82-2 (1982).

[90] DELANEY, E.G., MICKELSON, J.R., "USDOE decommissioning experience – Selected projects", presented at NEA Workshop, 22-24 Oct. 1984.

[91] KARKER, S., HOLMGREN, P., "Dismantling of the R-1 Reactor, Stockholm", 1982, International Decommissioning Symposium (Proc. Int. Symp. Seattle, 1982), CONF-821005, UNC Nuclear Industries, Richland, WA, Technical information Center, USDOE (1982).

[92] KITTINGER, W.D., et al., "Lessons learned in decommissioning the sodium reactor experiment", 1982 International Decommissioning Symposium (Proc. Int. Symp. Seattle, 1982), CONF-821005, UNC Nuclear Industries, Richland, WA, Technical Information Center, USDOE (1982).

[93] ABEL, K.H., et al., Radionuclide Distribution and Inventory at the Humboldt Bay Nuclear Plant, Batelle Pacific Northwest Labs, Richland, WA, Rep. PNL-4628 (1983).

[94] ABEL, K.H., et al., Radionuclide Distribution and Inventory at Dresden Nuclear Power Staion Unit No. 1, Batelle Pacific Northwest Labs, Richland, WA, Rep. PNL-4961.

[95] INTERNATIONAL ATOMIC ENERGY AGENCY, Data from a questionnaire to selected Member States (1985).

[96] Retech, Inc., P.O. Box 997, UKIAH, Ca.

[97] BECKERS, R.M., et al., "Remotely operated plasma torch: A tool for nuclear reactor dismantling", 94th Annual Winter Meeting of American Society of Mechanical Engineers, 11-15 Nov. 1973, CONF-731105-2.

[98] E.H. Wachs Company, Wheeling, IL.

[99] Onada Cement Co. Ltd, Tokyo, Japan.

[100] Sumituomo Cement Co. Ltd, Tokyo, Japan.

[101] GREGORY, A.R., Central Electricity Generating Board, UK (1985).

[102] BIRSS, I.R., GREGORY, A.R., "IKAEA and CEGB development programmes", Decommissioning of Radioactive Facilities (Proc. Seminar London, 7 Nov. 1984).

[103] MacDonald Air Tool Company, New Jersey, NJ, USA.

[104] 1984/85 UK Robots Industry Directory, British Robot Association, Kempston, Bedford, UK.

[105] KOHLER, G.W., Manipulator Type Book, Verlag Karl Thiemig, Munich (1981).

[106] STOIJKOVIC, Z., SALETIC, D., "Learning to recognize patterns by Belgrade hand prosthesis", Industrial Robots (Proc. 5th Int. Symp. Chicago, 1975).

[107] Cincinnati Milacron Inc., Cincinnati, OH, USA, The T^3 Industrial Robot.

[108] GRASP — A Graphical Robot Applications Simulation Package, BYG Systems Ltd, Nottingham, UK.

[109] WHITE, J.R., et al., Evaluation of Robotic Inspection Systems at Nuclear Power Plants, Missouri Univ., Columbia, Rep. NUREG-CR-3717 (Mar. 1984)

[110] Programmed and Remote Systems Corp., St. Paul, MN, USA, Pedestal, wall and crane mounted manipulators.

[111] Tale Operator Systems Corp., St. James, NY, USA.

[112] KUBAN, D., MARTIN, H., "An advanced remotely maintainable servomanipulator concept", Robotics and Remote Handling in Hostile Environments (Proc. ANS National Topical Meeting, Gatlinburg, 1984) 407.

[113] VERTUT, J., et al., "MA-23M contained servomanipulator with television cameras on PICA and PIADE telescopic supports with computerized integrated control", Remote System Technology (Proc. 28th Conf. Washington, DC, 1980), Vol. 2.

[114] Lamberton Robotics Limited, Coatbridge, Scotland.

[115] DEVRESSE, M., et al., "Vehicle for remote inspection in environments unhealthy or non-accessible for humans", Robotics and Remote Handling in the Nuclear Industry (Proc. Int. Conf. Toronto, 1984) Canadian Nuclear Society.

[116] CHESTER, C.V., Improved Robotic Equipment for Radiological Emergencies, Oak Ridge National Lab., TN, Rep. ORNL-6081 (1984).

[117] Da COSTA, L., et al., Review of Systems for Remotely Controlled Decommissioning Operations, Commission of the European Communities, Brussels, Rep. EUR (1985).

[118] General Electric Company, King of Prussia, PA, USA, Man-Mate 1600 Industrial Manipulators and Submersibles with Manipulators.

[119] NORGATE, G., "The future of artificial intelligence in nuclear plant maintenance", Robotics and Remote Handling in the Nuclear Industry (Proc. Int. Conf. Toronto, 1984), Canadian Nuclear Society.

[120] CONRATH, J.J., "Remotely controlled repair of piping at Douglas Point", Robotics and Remote Handling in the Nuclear Industry (Proc. Int. Conf. Toronto, 1984), Canadian Nuclear Society.

[121] OSANAI, M., "JPDR decommissioning program", 1982 International Decommissioning Symposium (Proc. Int. Symp. Seattle, Oct. 1982), CONF-821005, UNC Nuclear Industries, Richland, WA, Technical Information Center, USDOE (1982).

[122] LAWTON, H., "Decommissioning Windscale advanced gas-cooled reactor", Decommissioning of Radioactive Facilities (Proc. Seminar London, 7 Nov. 1984).

[123] KRIEGER, F., VIECENZ, H.J., "Remote dismantling of the pressure tube reactor from NPP Niederaichbach", Robotics and Remote Handling in the Nuclear Industry (Proc. Int. Conf. Toronto, 1984), Canadian Nuclear Society.

LIST OF PARTICIPANTS

De, P.L.
(Chairman)

Atomic Energy of Canada Limited,
CANDU Operations — Montreal,
2nd Floor, 1155 Metcalfe Street,
Montreal, Canada

Abel, E.

Engineering Projects Division,
UKAEA Atomic Energy Research Establishment,
Harwell, Didcot,
Oxfordshire OX11 0RA, United Kingdom

Bertini, A.

Ente Nazionale per l'Energia Atomica (ENEL/DPT),
Via G.B. Martini 3,
Piazza Verdi,
I-00198 Rome, Italy

Choudhary, R.S.

Rajasthan Atomic Power Station,
P.O. Anushakti, Via Kota,
Rajasthan, 323 303, India

Conti, M.

Comitato Nazionale per la Ricerca e per
  lo Sviluppo dell'Energia Nucleare e delle
  Energie Alternative (ENEA/PAS/SMAIMP),
CRE Casaccia,
C.P. 2400,
I-00100 Rome, Italy

Crégut, A.

CEA, Centre d'études nucléaires de la Vallée du Rhône,
B.P. 171, Marcoule,
F-30205 Bagnols-sur Cèze, France

Dekais, J.J.

S.A. Belgonucléaire,
Rue du Champs de Mars 25,
B-1050 Brussels, Belgium

Diefenbacher, W.

Kernforschungszentrum Karlsruhe GmbH,
Projektträger UB/SN,
Postfach 3640,
D-7500 Karlsruhe 1, Federal Republic of Germany

Fujishima, N.

Nuclear Safety Division,
Nuclear Safety Bureau,
Science and Technology Agency,
2-2-1 Kasumigaseki, Chiyoda-ku,
Tokyo, Japan

Gregory, A.R.

Nuclear Decommissioning Project,
Central Electricity Generating Board,
Barnett Way,
Gloucester GL4 7RS, United Kingdom

Heine, W.

Decommissioning Program,
Rockwell International,
Rockwell Hanford Operations,
Hanford, WA 99352, United States of America

Heinonen, J.

Reactor Laboratory,
Technical Research Centre of Finland (VTT),
SF-02150 Espoo 15, Finland

Huber, B.
(CEC)

Commission of the European Communities,
DG XII-SDM-1/85,
Rue de la Loi 200,
B-1049 Brussels, Belgium

Kirk, J.

UKAEA Atomic Energy Research Establishment,
Dounreay, Caithness KW14 7TZ, United Kingdom

Laraia, M.

Comitato Nazionale per la Ricerca e per
lo Sviluppo dell'Energia Nucleare e
delle Energie Alternative (ENEA/DISP),
Via Vitaliano Brancati 48,
I-00144 Rome, Italy

Leake, J.W.

UKAEA Atomic Energy Research Establishment,
Harwell, Didcot,
Oxfordshire OX11 0RA, United Kingdom

Leconte, M.

Division nucléaire,
Ateliers de constructions électriques de Charleroi,
Avenue Rousseau Marc,
B.P.4,
B-6000 Charleroi, Belgium

Liu, Zunqui

Bureau of Nuclear Fuels,
P.O. Box 2102-10,
Beijing, China

Lurie, R.

CEA, Centre d'études nucléaires de la Vallée du Rhône,
Institut de protection et de sûreté nucléaire,
B.P. 171, Marcoule,
F-30205 Bagnols-sur-Cèze, France

Motte, F.

Centre d'étude de l'énergie nucléaire (CEN/SCK),
Boeretang 200,
B-2400 Mol, Belgium

Niino, T.   Nuclear Construction Engineering Department,
                                 Hitachi Ltd,
                                 1-1, Saiwaicho 3 chome,
                                 Hitachi-shi 312, Ibaraki-ken,
                                 Japan

Obreja, Gh.   Technical Department,
                                 State Committee for Nuclear Energy,
                                 Magurele-Bucharest, Romania

Shimura, T.   Decommissioning Technique Section,
                                 Department of Development,
                                 Nuclear Power Engineering Test Center,
                                 No. 2 Akiyama Bldg,
                                 3-6-2 Toranomon,
                                 Minato-ku, Tokyo 105, Japan

Spencer, R.H.J.   UKAEA, Safety and Reliability Directorate,
                                 Culcheth, Risley,
                                 Warrington WA3 6AT, United Kingdom

Stearn, S.   H.M. Radiochemical Inspectorate,
                                 Department of the Environment,
                                 Room A5.31, Romney House,
                                 43, Marsham Street,
                                 London SWIP 3 PY, United Kingdom

Tomczak, W.K.   Radioactive Waste Disposal Department,
                                 Institute of Atomic Energy,
                                 05-400 Świerk, Poland

Välimäki, P.   Imatran Voima Oy,
                                 P.O. Box 138,
                                 SF-00101 Helsinki, Finland

Yin, Guowei   South-West Reactor Engineering,
                                 Research and Designing Institute,
                                 P.O. Box 291-(206),
                                 Chengdu, Sichuan, China

HOW TO ORDER IAEA PUBLICATIONS

An exclusive sales agent for IAEA publications, to whom all orders
and inquiries should be addressed, has been appointed
in the following country:

UNITED STATES OF AMERICA   Bernan Associates — UNIPUB, 10033-F King Highway, Lanham, MD 20706

In the following countries IAEA publications may be purchased from the
sales agents or booksellers listed or through your
major local booksellers. Payment can be made in local
currency or with UNESCO coupons.

ARGENTINA	Comisión Nacional de Energía Atómica, Avenida del Libertador 8250, RA-1429 Buenos Aires
AUSTRALIA	Hunter Publications, 58 A Gipps Street, Collingwood, Victoria 3066
BELGIUM	Service Courrier UNESCO, 202, Avenue du Roi, B-1060 Brussels
CHILE	Comisión Chilena de Energía Nuclear, Venta de Publicaciones, Amunategui 95, Casilla 188-D, Santiago
CHINA	IAEA Publications in Chinese: China Nuclear Energy Industry Corporation, Translation Section, P.O. Box 2103, Beijing IAEA Publications other than in Chinese: China National Publications Import & Export Corporation, Deutsche Abteilung, P.O. Box 88, Beijing
CZECHOSLOVAKIA	S.N.T.L., Mikulandska 4, CS-116 86 Prague 1 Alfa, Publishers, Hurbanovo námestie 3, CS-815 89 Bratislava
FRANCE	Office International de Documentation et Librairie, 48, rue Gay-Lussac, F-75240 Paris Cedex 05
HUNGARY	Kultura, Hungarian Foreign Trading Company, P.O. Box 149, H-1389 Budapest 62
INDIA	Oxford Book and Stationery Co., 17, Park Street, Calcutta-700 016 Oxford Book and Stationery Co., Scindia House, New Delhi-110 001
ISRAEL	Heiliger and Co., Ltd, Scientific and Medical Books, 3, Nathan Strauss Street, Jerusalem 94227
ITALY	Libreria Scientifica, Dott. Lucio de Biasio "aeiou", Via Meravigli 16, I-20123 Milan
JAPAN	Maruzen Company, Ltd, P.O. Box 5050, 100-31 Tokyo International
NETHERLANDS	Martinus Nijhoff B.V., Booksellers, Lange Voorhout 9-11, P.O. Box 269, NL-2501 The Hague
PAKISTAN	Mirza Book Agency, 65, Shahrah Quaid-e-Azam, P.O. Box 729, Lahore 3
POLAND	Ars Polona-Ruch, Centrala Handlu Zagranicznego, Krakowskie Przedmiescie 7, PL-00-068 Warsaw
ROMANIA	Ilexim, P O. Box 136-137, Bucharest
SOUTH AFRICA	Van Schaik Bookstore (Pty) Ltd, P.O. Box 724, Pretoria 0001
SPAIN	Díaz de Santos, Lagasca 95, E-28006 Madrid Díaz de Santos, Balmes 417, E-08022 Barcelona
SWEDEN	AB Fritzes Kungl. Hovbokhandel, Fredsgatan 2, P.O. Box 16356, S-103 27 Stockholm
UNITED KINGDOM	Her Majesty's Stationery Office, Publications Centre, Agency Section, 51 Nine Elms Lane, London SW8 5DR
USSR	Mezhdunarodnaya Kniga, Smolenskaya-Sennaya 32-34, Moscow G-200
YUGOSLAVIA	Jugoslovenska Knjiga, Terazije 27, P.O. Box 36, YU-11001 Belgrade

---

Orders from countries where sales agents have not yet been appointed and
requests for information should be addressed directly to:

Division of Publications
International Atomic Energy Agency
Wagramerstrasse 5, P.O. Box 100, A-1400 Vienna, Austria

86-03880